CELL BIOLOGY RESEARCH PROGRESS

LIPID BILAYERS

PROPERTIES, BEHAVIOR AND INTERACTIONS

CELL BIOLOGY RESEARCH PROGRESS

Additional books and e-books in this series can be found
on Nova's website under the Series tab.

CELL BIOLOGY RESEARCH PROGRESS

LIPID BILAYERS

PROPERTIES, BEHAVIOR AND INTERACTIONS

MOHAMMAD ASHRAFUZZAMAN
EDITOR

Copyright © 2019 by Nova Science Publishers, Inc.

All rights reserved. No part of this book may be reproduced, stored in a retrieval system or transmitted in any form or by any means: electronic, electrostatic, magnetic, tape, mechanical photocopying, recording or otherwise without the written permission of the Publisher.

We have partnered with Copyright Clearance Center to make it easy for you to obtain permissions to reuse content from this publication. Simply navigate to this publication's page on Nova's website and locate the "Get Permission" button below the title description. This button is linked directly to the title's permission page on copyright.com. Alternatively, you can visit copyright.com and search by title, ISBN, or ISSN.

For further questions about using the service on copyright.com, please contact:
Copyright Clearance Center
Phone: +1-(978) 750-8400   Fax: +1-(978) 750-4470   E-mail: info@copyright.com

NOTICE TO THE READER

The Publisher has taken reasonable care in the preparation of this book, but makes no expressed or implied warranty of any kind and assumes no responsibility for any errors or omissions. No liability is assumed for incidental or consequential damages in connection with or arising out of information contained in this book. The Publisher shall not be liable for any special, consequential, or exemplary damages resulting, in whole or in part, from the readers' use of, or reliance upon, this material. Any parts of this book based on government reports are so indicated and copyright is claimed for those parts to the extent applicable to compilations of such works.

Independent verification should be sought for any data, advice or recommendations contained in this book. In addition, no responsibility is assumed by the Publisher for any injury and/or damage to persons or property arising from any methods, products, instructions, ideas or otherwise contained in this publication.

This publication is designed to provide accurate and authoritative information with regard to the subject matter covered herein. It is sold with the clear understanding that the Publisher is not engaged in rendering legal or any other professional services. If legal or any other expert assistance is required, the services of a competent person should be sought. FROM A DECLARATION OF PARTICIPANTS JOINTLY ADOPTED BY A COMMITTEE OF THE AMERICAN BAR ASSOCIATION AND A COMMITTEE OF PUBLISHERS.

Additional color graphics may be available in the e-book version of this book.

Library of Congress Cataloging-in-Publication Data

ISBN: 978-1-53616-392-6
Library of Congress Control Number:2019950409

Published by Nova Science Publishers, Inc. † New York

CONTENTS

Preface vii

Acknowledgments xi

Chapter 1   Fundamentals of Plasma, Nuclear
and Mitochondrial Lipid Membranes   1
Mohammad Ashrafuzzaman

Chapter 2   Mechanical Stresses in the Lipid Bilayer
of Erythrocyte Membranes   43
Pavel V. Mokrushnikov

Chapter 3   The Role of the Lipid Bilayer in the Erythrocyte
Membrane Structural Changes   93
Ivana Pajic-Lijakovic and Milan Milivojevic

Chapter 4   Microbial-Derived Bioactive Lipopeptides:
Pore-Forming Metabolites in Lipid Bilayers   123
*Yulissa Ochoa, Mariana Bernal
and Daniel Balleza*

Chapter 5	Colchicine Induced Ion Channel Formation into Membranes as a Mechanism behind Chemotherapy Drug Cytotoxicity of Cancer Cells *A. A. Alqarni, S. Zargar and Md. Ashrafuzzaman*	**151**
Chapter 6	Molecular Stability Analysis of Reverse Micelles Role in Breast Cancer Drug Delivery and It's Dynamics and Simulation Studies *Dhivya Shanmugarajan, N. Premjanu, Ganesh Munuswamy, Lakshmi Jayasri Akkiraju, Sushil Kumar Middha and Sureshkumar Chinanga*	**189**
About the Editor		**215**
Index		**217**
Related Nova Publication		**229**

PREFACE

Lipid bilayers are major structural parts of cell membranes. The membrane consists of lipids, membrane proteins, cholesterols, hydrocarbons, etc. Lipid bilayer creates a hydrophobic/hydrophilic boundary which helps shape cellular structures as well as maintain controlled transport of materials and information across the boundary. Understanding lipid bilayer requires biological, physiological, biophysical and biochemical addresses.

Over the past four decades, there has been a growing dialogue between biologists and an ever-increasing number of physicists, chemists, mathematicians, and engineers. The reason for this mutual interest is the modern biology's pre-eminent role on the front lines of scientific research.

The connection between the physical and biomedical sciences has been developing rapidly over the past few decades, especially since the groundbreaking discoveries in molecular genetics. There is a need for a continuing dialogue and a cross-fertilization between life sciences and physical sciences. As a result of the naturally interdisciplinary nature of the life sciences, numerous new border areas are being created and developed. Therefore, disciplines such as mathematical biology, biophysics, computational biology, Bioinformatics, biostatistics, biological physics, theoretical biology, biochemistry, and biomedical engineering have been undergoing exponential growth.

The diversity present in biological systems is a simple result of the multitude of possible combinations of the finite number of structural elements. The functioning of biological systems should also follow from this complexity with the specific biological organization of complex biochemical and biomolecular systems providing specific functions while they continue to be governed by fundamental physical laws. The principle of biological complexity begetting function is familiar to physicists and has often been referred to as an emergent phenomenon. It is a characteristic feature of atomic systems to display new emergent properties as they become more complex. This property is at the core of living organisms that acquire new functional features as their structural complexity grows with size from single-cell organisms to multi-cellular organisms.

This hierarchical, interconnected, but coherently operating system of biological systems that sustains life poses a great scientific challenge not only to understand how its pieces work but how the whole is organized internally to achieve specific functional advantages. Our quest to understand structures of biological systems having specific types of chemical processes active using chemical analysis, physical laws, and engineering principles is greatly aided by the rapid development of sophisticated experimental techniques that physics and technology have supplied for the use of biologists. Therefore, biologists, physicists, chemists, and engineers may find clear scientific reasons to cross-examine their ideas and approaches while exploring biological systems. Some of the most prominent examples of techniques, used independently or in collaboration in all these core branches of scientific research, are listed below:

- Light microscope (resolution: 400-600 nm) with various modern upgrades such as confocal, phase contrast or cryomicroscopy.
- Electron microscope (10-100 nm)
- Neutron scattering (1-10 Å)
- X-ray crystallography (1 Å)
- Patch-clamp electrophysiology
- STM, AFM, TEM
- NMR, MRI. fMRI

- Fluorescence spectroscopy
- Microwave absorption
- Laser light scattering
- Synchrotron radiation scattering
- Laser tweezers
- Bioinformatics,

etc.

This book is intended to provide a broad overview of an important biological system 'cell membrane'. The cell is the powerhouse where processes of life are controlled. The cell membrane is the cell's boundary that determines most of the cell-based uptakes of materials, exchange of information between both sides and ensures helping vital biological processes to continue. We have focused specifically on an understanding of various aspects of membrane bilayers. The book is focused on a detailed description of the diverse mechanisms and phenomena associated with membranes. Lipid bilayers exist in various parts of the cell membrane, namely, across the plasma membrane, mitochondrial membrane, and nuclear membrane. While exploring lipid bilayers we shall, therefore, need to consider structures and functions of various sections of biological cells. As it is an edited book, we have got a group of contributors writing chapters focusing on different aspects. Each chapter may be found to present an individual topic and elaborate on a specific problem. But all chapters altogether have covered most of the basic aspects relevant to the title of the book. Eminent scientists from all over the world have contributed.

General membrane phenomena, mechanisms, and other properties will be discussed in chapter 1. Here we have detailed the diversity of membrane structures and functions among various membranes, namely, plasma, mitochondrial and nuclear membranes. This chapter will serve the introductory materials for the book. This will be followed by chapters written by various groups with versatile expertise. Basic understanding of cell membrane phenomena and applied biological and biophysical address of lipid bilayer perspectives determining the cellular structures and functions and related applications in drug discovery, delivery, etc. have been covered.

Besides theoretical analysis and modeling of lipid bilayer structures, a bulk amount of experimental findings have been presented to the current topical issues. The readers will find this edited book to present the latest information about the topics covered here and provide clear guidelines about future trends. The editor and authors hope that this monograph will be found of use as a source of valuable information and conceptual inspiration to students and expert researchers.

ACKNOWLEDGMENTS

I am thankful to all contributing authors. All of them have in-depth knowledge about various aspects of lipid bilayers. Completion of this book would be impossible without using a bulk of experimental and theoretical data from publications of various other authors (all are quoted in references). It's a great pleasure to thank all of them for their excellent contributions to the field. Hundreds of discussions with colleagues, academic friends, research group members, and students helped to shape ideas while raising the ideas, choosing the chapters and ordering them during the last two years. Assistance and encouragement provided by the staff members at Nova Science are thankfully acknowledged. Imtihan Ahmed of IBM, Markham, Canada and Emma Mannan of Alberta University, Edmonton, Canada are thankfully acknowledged for their direct help in editing all chapters. I especially value the emotional support given by my wife Anwara. Anwara has been always on trips between my Canada residence and Saudi workstation to provide much needed social supports in my every academic activity.

Mohammad Ashrafuzzaman
Riyadh, Saudi Arabia, July 2019

In: Lipid Bilayers
Editor: Mohammad Ashrafuzzaman

ISBN: 978-1-53616-392-6
© 2019 Nova Science Publishers, Inc.

Chapter 1

FUNDAMENTALS OF PLASMA, NUCLEAR AND MITOCHONDRIAL LIPID MEMBRANES

Mohammad Ashrafuzzaman[*]
Department of Biochemistry, College of Science,
King Saud University, Riyadh, Saudi Arabia

ABSTRACT

Lipid bilayers participate in constructing important cellular structures, maintaining phenomenological transport properties, and regulating trans-bilayer communication. Biological cells' plasma, mitochondrial and nuclear membranes appear with different structures and functions. Plasma membrane bilayers help cells primarily to maintain their geometrical surfaces' dynamic structures and ensure regulating communication between extracellular and intracellular physiological environments. Both nuclear and mitochondrial membranes consist of two phospholipid (double lipid) bilayers. Nuclear lipid bilayers play crucial roles in ensuring correct genetic environment originated inside nucleus. Mitochondrial lipid bilayers are involved in transport mechanisms delivering biomolecules that are especially responsible to maintain cellular health. All these lipid bilayers continuously deal with naturally inbuilt physical, physiological

---

[*] Corresponding Author's Email: mashrafuzzaman@ksu.edu.sa, ashrafuz@ualberta.ca.

and biological phenomena that determine crucial defining aspects of cell in general and life in particular. This chapter will mainly focus at understanding the biological organization of all these three distinguishable lipid bilayers, as well as addressing their physical roles that determine versatile cell based physiological signaling pathways and functions.

Cell membranes make outer layers of cells. They participate in maintaining shapes of the cells as well as ensuring transport of nutrients between outer and inner regions of cells. Cell membranes consists of various materials, among them lipids and proteins (referred as membrane proteins), cholesterol, etc. are especially mentionable. The lipids usually organize themselves to construct lipid layers in cell membranes. Lipid bilayers participate in constructing important cellular structures, maintaining phenomenological transport properties, and regulating trans-bilayer communication. Biological cells' plasma, mitochondrial and nuclear membranes appear with different structures and functions. Although a lot of identical lipids and versatile membrane proteins are found in all these three membranes their organizations and functions are distinguishable. Plasma membrane bilayers help cells primarily to maintain their geometrical surfaces' dynamic structures and ensure regulating communication between extracellular and intracellular physiological environments (Nicholsson and Springer, 1971; Springer, 2014). Both nuclear and mitochondrial membranes consist of two phospholipid (double lipid) bilayers. Nuclear lipid bilayers play crucial roles in ensuring correct genetic environment originated inside nucleus. Mitochondrial lipid bilayers are involved in transport mechanisms delivering biomolecules that are especially responsible to maintain cellular health. All these lipid bilayers continuously deal with naturally inbuilt physical, physiological and biological phenomena that determine crucial defining aspects of cell in general and life in particular. This chapter will mainly focus at understanding the biological organization of all these three distinguishable lipid bilayers, as well as addressing their physical roles that determine versatile cell based physiological signaling pathways and functions.

1.1. Plasma Membrane

Cell membrane is a thick layer surrounding any biological cell. It contains two major classes of chemical compositions, namely (i) a lipid bilayer, and (ii) various membrane proteins. Besides, there are hydrocarbons. Varieties of phospholipids, glycolipids, proteins, cholesterol and hydrocarbons help a cell's plasma membrane to get a biologically distinguishable geometric structure. The five decades old 'fluid-mosaic membrane model (F-MMM)' of Nicholson and Springer demonstrated a rather complete membrane structure that is still largely valid in fundamental level (see (Singer and Nicolson, 1972) or revised (Nicolson et al., 1977; Nicolson, 1976)). A lot of pinpointed studies during last five decades discovered many fundamental aspects. The understanding of membrane now involves addressing through biological, physiological, biochemical and biophysical concepts and parameters. Combining all these versatile components and information the plasma membrane model has recently been summarized by F-MMM co-discoverer Nicolson, see Figure 1.1 (Nicolson, 2014), for details see book 'Nanoscale Biophysics of the Cell' (Ashrafuzzaman, 2018). Nicolson considered the data published mainly during 1972-2014, presented an analysis and cleared some of the misconceptions inherent in the 1970s era models. We also encourage readers to try to understand the membrane surface mechanical properties through our model diagram sketch (by Imtihan Ahmed and Mohammad Ashrafuzzaman), as presented in Figure 1.2, for details see book 'Membrane Biophysics' (Ashrafuzzaman and Tuszynski, 2012).

The Figures 1.1 and 1.2 include the presence of different integral proteins, glycoproteins, lipids and oligosaccharides represented by different colors. The membrane has been peeled-up to view the inner membrane surface cytoskeletal fencing that restricts the lateral diffusion of some, but not all trans-membrane glycoproteins. Other important lateral diffusion restriction mechanisms are also represented, such as lipid domains, integral membrane glycoprotein complex formation (seen in the membrane cut-away), polysaccharide–glycoprotein associations (at the far top left, Figure 1.1) and direct or indirect attachment of inner surface membrane domains to cytoskeletal elements (at lower left, Figure 1.1). Although the Figure 1.1 suggests some possible integral MP and lipid mobility restraint mechanisms, it does not accurately present the sizes or structures of integral MPs, cytoskeletal structures, polysaccharides, lipids, submicro- or nano-sized domains or membrane-associated cytoskeletal structures or their crowding in the membrane.

In this latest model (figures 1.1 and 1.2), inclusion of the following three rather important information are especially mentionable:

1) Integral membrane protein lateral movements (for details, see ref. (Jacobson et al., 1995))
2) Lipid domains around integral membrane proteins and glycoproteins (for details, see ref. (Escribá et al., 2008))

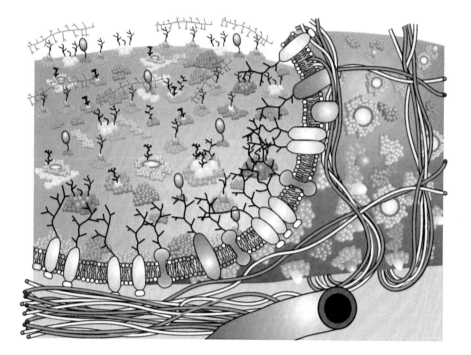

Figure 1.1. An updated F-MMM that contains information on membrane domain structures and membrane-associated cytoskeletal and extracellular structures (see ref. (Nicolson, 2014)).

Figure 1.2. A simplified membrane's core structure has been presented here. This diagram was sketched by Imtihan Ahmed and Mohammad Ashrafuzzaman for book 'Membrane Biophysics' (Ashrafuzzaman and Tuszynski, 2012). The three dimensional structure shows that the 3-5 nanometer thick membrane bilayer surface is not smooth. The mechanical properties (especially the elasticity) of membrane surface is reflected through change of bilayer thickness at random locations which may evolve due to mainly the mismatches between various integral membrane proteins coupled to bilayer across the bilayer thickness. Near lipid headgroups (light blue color) are seen the cholesterol, residing in the hydrophobic region. Globular proteins (in red) are shown to reside across the membrane. Alpha helical proteins have both hydrophilic and hydrophobic parts. Membrane bending, membrane thickness change, etc. are also schematically diagrammed here which are consistent with the real membrane surrounding a cell. For simplicity, we did not schematize the presence of different ion channels, or other complicated membrane protein structures, membrane domain structures and membrane-associated cytoskeletal and extracellular structures, as shown in Figure 1.1.

3) Membrane surface appears unsmooth due to mechanical properties of the lipid bilayer membrane. The bilayer thickness change occurs as a result of the mismatch caused by various integral membrane proteins while locally coupling with the bilayer. The locations where lipid bilayer does not contain integral proteins may show equilibrium bilayer thickness (for details, see ref. (Ashrafuzzaman and Tuszynski, 2012)). In a recent book 'Nanoscale Biophysics of the Cell' the pre and post F-MMM (see (Singer and Nicolson, 1972) or revised (Nicolson, 1976)) developments have been rigorously explained (Ashrafuzzaman, 2018). We shall therefore not elaborate on this issue and invite readers to go through this monograph (Ashrafuzzaman, 2018).

1.1.1. Diversity of Lipids in Plasma Lipid Bilayers

Lipid bilayers contain various classes of lipids performing different kinds of activities that help cells perform general functions and in special cases according to the specific needs of specific organelles and tissues. Lipids in membrane are classified as phospholipids, sphingolipids or sterols. Some of these lipids are attached to carbohydrates classifying them as glycolipids. The carbohydrate moiety could be one to several molecules of galactose or glucose. It is generally understood that the primary role of lipids in cellular function is to form the permeability barrier of cells and subcellular organelles in the form of a lipid bilayer.

Dowhan and Bogdanov summarized in a chapter outlining the diversity in structure, chemical properties, and physical properties of lipids, and presented various genetic approaches available for studying lipid function *in vivo*, and described how the physical and chemical properties of lipids relate to their multiple functions in living systems (see ref. (Dowhan and Bogdanov, 2002)). The review has detailed on how the roles lipids playing in cellular processes may be as diverse as the chemical structures of lipids. A single phospholipid component can form a sealed bilayer vesicle in solution, but a diversity of lipid structure and physical properties is necessary for filling the broad range of roles that lipids play in cells. The diverse functions of lipids are made possible by a family of low molecular weight molecules that are physically fluid and deformable to enable interaction in a flexible and specific manner with macromolecules.

Van Meer and co-investigators, in one of their important articles, focused quite in details on various structural lipids, summarized in Figure 1.3 (see ref. (Meer et al., 2008)). In eukaryotic membranes the major lipids are the glycerophospholipids: phosphatidylcholine (PC), phosphatidylethanolamine (PE), phosphatidylserine (PS), phosphatidylinositol (PI) and phosphatidic acid (PA).

The lipids' hydrophobic portion is a diacylglycerol (DAG), which contains saturated or *cis*-unsaturated fatty acyl chains of varying lengths. PC accounts for >50% of the phospholipids in most eukaryotic membranes. It self-organizes spontaneously as a planar bilayer in which each PC has a nearly cylindrical molecular geometry, with the lipidic tails facing each other and the polar headgroups interfacing with the aqueous phase. Most PC molecules have one *cis*-unsaturated fatty acyl chain, which renders them fluid at room temperature. PE has conical molecular geometry due to having relatively small size polar headgroup. The inclusion of PE in PC bilayers imposes a curvature stress onto the membrane, which is used for budding, fission and fusion (Marsh, 2007).

There are a few non-bilayer lipids, e.g., PE and cardiolipin (CL) which may also be used to accommodate membrane proteins and modulate their activities (March, 2007; Dowhan and Bogdanov, 2002; Ashrafuzzaman and Tuszynski, 2012). Various lipids' asymmetric distribution between the two bilayer leaflets, namely inner bilayer and outer bilayer leaflets, are found to contribute substantially into curvature stresses in biomembranes. Curvature stress leads to local change of energetics. A model is recently developed that connects molecular interactions to curvature stress, and which explains the role of local composition, e.g., sphingomyelin and cholesterol in plasma membrane (Sodt et al., 2016).

A theoretical analysis of simulations of lipid mixtures demonstrates that lipid spontaneous curvatures are frequently non-additive. Changes in lipid composition can have dramatic effects, including changing the sign of the spontaneous curvature, namely, positive or negative curvatures. In this theoretical demonstration, cholesterol is shown to lower the number of effective Kuhn segments of saturated acyl chains, reducing lateral pressure below the neutral surface of bending and favoring positive curvature. The effect is not observed for unsaturated (exible) acyl chains. Likewise, hydrogen bonding between sphingomyelin lipids leads to positive curvature, but only at sufficient concentration, below which the lipid prefers negative curvature.

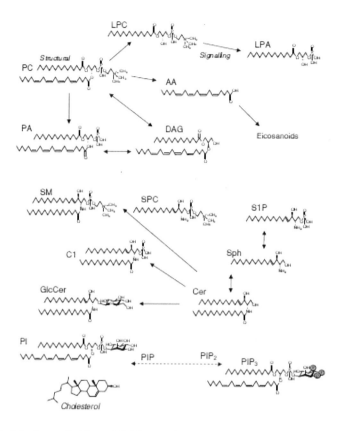

Figure 1.3. Membrane lipids and lipid second messengers. The main eukaryotic membrane lipids are the glycerophospholipids PC. Their DAG backbone carries a phosphate PA and either a choline (forming PC), ethanolamine (forming PE), serine (forming PS), or inositol (forming PI). The prototypical phospholipid, dipalmitoyl-PtdCho, exhibits nearly cylindrical molecular geometry with a cross-sectional surface area of 64 Å^2 and a head-to-tail length of 19 Å (Kucerka et al., 2006). The phosphosphingolipid sphingomyelin (SM) and the glycosphingolipid glucosylceramide (GlcCer) have a ceramide (Cer) backbone, consisting of a sphingoid base (such as sphingosine; Sph), which is an amide linked to a fatty acid. Yeast sphingolipids carry a C26 fatty acid and have phosphoinositol-X substituents that contain additional mannoses and phosphates. Breakdown products of membrane lipids serve as lipid second messengers. The glycerolipid-derived signalling molecules include lysoPtdCho (LPC), lysoPA (LPA), PA and DAG. The sphingolipid-derived signalling molecules include sphingosylphosphorylcholine (SPC), Sph, sphingosine-1-phosphate (S1P), Cer-1-phosphate (C1P) and Cer. Arachidonic acid (AA) yields the signalling eicosanoids and endocannabinoids. The various phosphorylated PtdIns molecules (also known as the phosphoinositides) mark cellular membranes and recruit cytosolic proteins. They are interconverted by the actions of kinases and phosphatases. Figure is redrawn in light of figure in ref. (Meer et al., 2008).

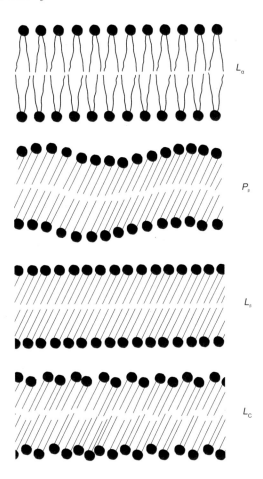

Figure 1.4. Top to bottom structures schematically represent fluid (L_α), ripple (P_β), gel (L_β) and pseudocystalline or subgel (L_c) states in the organization of the lamellar bilayer phases of PC (Jain, 1988; Meer et al., 2008). Each lipid is schematically presented with a head group (black sphere) and two tails. Various lamellar phases correspond to different organizations and orientations of the lipids.

The sphingolipids constitute another class of structural lipids. Their hydrophobic backbone is ceramide. The major sphingolipids in mammalian cells are sphingomyelin and the glycosphingolipids (GSLs), which contain mono-, di- or oligosaccharides based on glucosylceramide (GlcCer) and sometimes galactosylceramide (GalCer) (Meer et al., 2002). Gangliosides are GSLs with terminal sialic acids. Sphingolipids have saturated (or *trans*-unsaturated) tails so are able to form taller and narrower cylinders than PC lipids of the same chain length and pack more tightly, adopting the solid 'gel' or s_o phase (see various lipid phases modeled in Figure 1.4); they are also fluidized by sterols. A rigorous analysis on lipid phase properties has been made in our book 'Membrane Biophysics' (Ashrafuzzaman and Tuszynski, 2012). I shall avoid going into details and invite readers to go through this book for an in depth understanding of the lipid phase induced membrane behavior. However, I wish to present here a summary of materials and chemical conditions concerning lipids in membrane structure as we find in our analysis presented here and as briefly listed in ref. (Deamer, 2017):

- Phosphate is a primary anionic component of most membrane lipids. Sulfate and carboxylate groups are also present on certain lipids.
- Two fatty acids are attached to a glycerol by ester or ether bonds.
- The hydrocarbon chains must be in a fluid state at temperatures ranging from 0 to 100 °C depending on the organism's environment.
- The hydrocarbon chain lengths must be sufficiently long to be stable as bilayers, and also to maintain a permeability barrier to ionic and polar solutes. Fatty acids as short as 10 carbons can assemble into fragile membranes, but typical chain lengths of eukaryotic membrane phospholipids are in the range of 16 to 18 carbons.
- Biological membranes are not composed of pure phospholipids, but instead are mixtures of several phospholipid species, often with a sterol admixture such as cholesterol.
- The anionic groups of phospholipids and amino acids strongly interact with divalent cations, particularly calcium (Ca^{2+}), so living cells exclude calcium by active outward transport.
- Protons also interact with carboxylate, phosphate and amine groups on lipids and proteins. The protonation state strongly affects lipid properties, so cells maintain intracellular pH near neutrality (pH ~7) by actively pumping protons across their membranes.
- Lipid matrices between intracellular and extracellular lipid monolayers vary significantly.

1.1.2. Diversity of Proteins in Plasma Lipid Bilayers

Two types of proteins, namely, carrier and transmembrane proteins are contained in membranes. Both are engaged in helping membrane to (firstly) get constructed and (secondly) transport and exchange materials across the barrier following standard biophysical and biochemical principles and mechanisms.

Carrier proteins are integral components of fatty acid synthases, polyketide synthases, and nonribosomal peptide synthetases and play critical roles in the biosynthesis of fatty acids, polyketides, and nonribosomal peptides (for details see ref. (Lin et al., 2012)). Carrier proteins are engaged in binding solutes such as small organic molecules, inorganic ions, etc. on one side of the membrane and delivering them to the other side while experiencing changes in their conformation. Carrier proteins are often highly selective regarding the transported solutes and the transport mechanisms may be passive or active. The passive transport works down the electrochemical gradient. Specialized carrier proteins are engaged to facilitate the diffusion of specific molecules that move down a concentration gradient. Carrier proteins that transport molecules against a gradient can be directly coupled to hydrolysis of ATP, a process that provides the energy to drive the uphill, active and unidirectional transport process. The sodium/potassium ATPase (Na^+/K^+-ATPase) antiporter is an example of active transport. This active transport pump is located in the plasma membrane of every cell. This antiporter pumps 3 Na^+ out and 2 K^+ in for every ATP hydrolyzed (Pelley, 2007).

Carrier proteins are found in the plasma membranes, as well as in membranes of organelles like mitochondria (inner membranes) and chloroplasts. These proteins help them maintain specific biochemistry and biophysical processes. We shall avoid elaborating on this topic, but there are a lot of resources (text books and published articles) where the readers may enrich their knowledge.

Membrane proteins are often classified as per their roles, e.g.,

- channel protein which allows charged and polar particles through the membrane,
- carrier proteins which force charged and polar particles through the membrane (uses ATP),
- receptor proteins that receive chemical signals from outside a cell,
- cell regulation proteins which help a *cell* communicate with its environment,
- enzymatic proteins that act as catalysts and help in complex reactions, etc.

Transmembrane proteins mostly extend across the bilayer as

1) a single α helix,
2) as multiple α helices, or
3) as a rolled-up β sheet (a β barrel).

Some of these "single-pass" and "multipass" proteins have a covalently attached fatty acid chain inserted in the cytosolic lipid monolayer. Other membrane proteins are exposed at only one side of the membrane.

1) Some of these are anchored to the cytosolic surface by an amphipathic α helix that partitions into the cytosolic monolayer of the lipid bilayer through the hydrophobic face of the helix.
2) Others are attached to the bilayer solely by a covalently attached lipid chain—either a fatty acid chain or a prenyl group—in the cytosolic monolayer or,
3) via an oligosaccharide linker, to phosphatidylinositol in the noncytosolic monolayer.

Many proteins are attached to the membrane only by noncovalent interactions with other membrane proteins. For details see ref. (Alberts et al., 2014).

Fundamentals of Plasma Nuclear and Mitochondrial Lipid ...   13

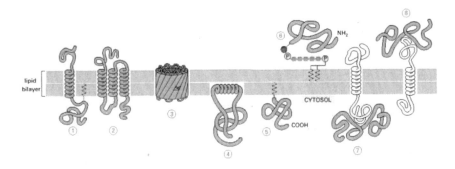

Figure 1.5. Membrane proteins associate with the lipid bilayer in various ways. This old sketch is still so correct, taken from ref. (Alberts et al., 2014).

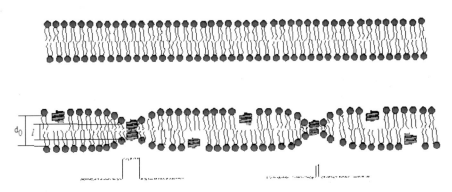

Figure 1.6. Gramicidin A (gA) channels deform lipid bilayer's resting thickness. While single (no channels are formed) gA monomers are monolayer peptides. With channel formation bilayer conducts a current pulse with an average pulse width (lifetime ~ ms) and height (channel current ~ pA). Two types of monomers have demonstrated formation of two different channels (Ashrafuzzaman and Tuszynski, 2011; 2012).

Peptides, depending on their size, structure, and other relevant biophysical properties, etc. may bind with lipid monolayers (either intracellular or extracellular) or span across the lipid bilayer due to physiological demands. These molecules are on continuous motion inside membrane and also participate in constructing specialized structures occasionally, e.g., pores or channels, defects, etc. Here we present the scenario using two peptides, namely gramicidin A (gA) and alamethicin (Alm) in lipid bilayers (see Figures 1.6 and 1.7).

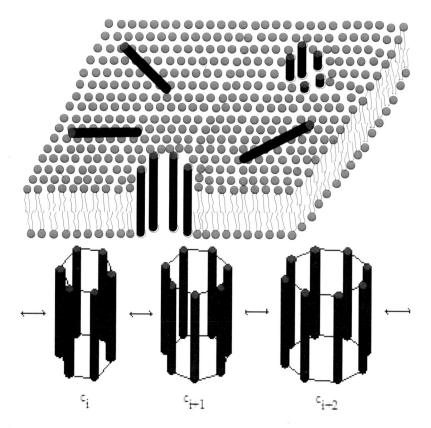

Figure 1.7. Alamethicin channels are 'barrel stave' type (Huang, 1986), constructed from multiple alamethicin monomers which are modeled here as cylindrical rods. Transition between different alamethicin conductance pores by addition/release of monomer(s) from/to the surrounding space is shown in lower panel. Here alamethicin monomer movement may happen in both transmembrane associated and monolayer conditions (Ashrafuzzaman and Tuszynski, 2011; 2012).

1.2. Electrical State of Membrane

Capacitance measurements of a cell's insulating compartment such as lipid membranes' hydrophobic cores and current measurements across membrane based physiological barriers provide crucial low dimension scale biological and chemical information on cell constructs.

We have discussed, quite in detail, the electrical, capacitive and related issues of cell membrane in our previous books 'Membrane Biophysics' (Ashrafuzzaman and Tuszynski, 2012) and Nanoscale Biophysics of the Cell (Ashrafuzzaman, 2018). The readers may consider reading these two books to understand the deep insights.

1.2.1. Membrane Potential

Membrane potential is the electrical potential difference between the interior and exterior of a cell. If the potential of the region just outside the membrane is V_o and the potential of the region just inside the cell near the membrane is V_i the membrane potential of the cell is V_i-V_o. Using traditional definition of electrical potential we can also define the membrane potential as the energy required to transfer unit charge from the exterior to the interior of a cell across the membrane. For example, if transfer of Q coulomb charge from exterior to interior of a cell requires energy of W joules the potential difference, hence the membrane potential of the cell will be W/Q volts.

Both cellular interior and exterior regions exist with electrical conditions represented by electrical potentials. Electrical potential of both regions depend mainly on what are the constituents making the regions. The fluids on both sides of the mainly lipid membrane contain high concentrations of various ions – both cations and anions. Among the cations sodium (Na^+), potassium (K^+) and calcium (Ca^{+2}) are mentionable while chloride (Cl^-) is the important anion. Although both cations and anions exist in both interior and exterior regions of a cell the concentrations of sodium and chloride ions in the exterior is higher than that in the interior. Similarly, potassium ions exist in higher concentration in the interior region than the exterior region of a cell. The interior region importantly exists with the giant presence of protein anions.

Due to the differences in charge types and concentrations between intracellular and extracellular regions they exist with different potential conditions, as a result membrane exists in an electric field which accounts for the membrane potential. The membrane therefore plays a role of cell's battery by providing a continuous source of electrical energy originated from the potential imbalance between intracellular and extracellular regions. This source of electrical energy also plays important roles in regulating many cellular processes like transmitting signals between different parts of a cell, exciting ion channels across the membrane, etc. As the concentrations of each of the ions in the intracellular and extracellular regions are different, there always exists a concentration gradient for each ion across the cell membrane. This gradient causes a tendency for the ions to cross through the membrane. Potassium ion tries to move from intracellular to the extracellular region while sodium and chloride try to flow in the opposite direction. The natural tendency of the movement of charges across membrane causes changes in the membrane resting potential. Similarly, the changes in membrane resting potential due to natural or due to artificial stimulus drive charges across the membrane. These rather slow dynamical processes are keys to normal functioning of cellular processes.

1.2.2. Resting Potential

Resting potential is simply the potential of a membrane's interior in absence of any excitation. This is in fact the membrane potential of non-excitable cell or the membrane potential of an excitable cell in absence of any excitations. Besides the uneven distribution of other charges as explained earlier about 10 times higher Na^+ concentration on the outside and 20 times higher K^+ concentration on the inside of a membrane cause altogether the huge charge density gradient. As a result, in condition of rest or in absence of any excitation the cell membrane is polarized maintaining an effective negative interior charge which accounts for a negative interior resting potential of the order of about -70 mV (see Figure 1.8).

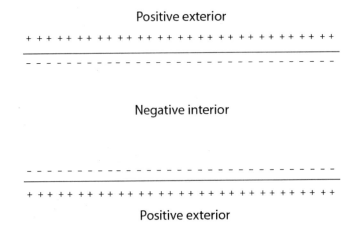

Figure 1.8. About -70 mV potential in the membrane interior region relative to the membrane exterior region is a general electrical condition found due to a resultant negatively charged interior and positively charged exterior membrane regions.

The chemical gradient across the membrane in rest causes resting potential to get built. ATP powered ion pumps or ion transporters play crucial roles. In an animal cell plasma membrane sodium-potassium pumps (Na^+ or K^+-ATPase) help to build sodium and potassium gradients across the membrane. The resting potential may also be altered due to the change of acidic environment across the cells. E.g., in cancer cell membrane due to extra acidic condition the resting potential may alter considerably.

1.2.3. Membrane as a Capacitor

A cell membrane separates charges on both sides of it. The inner core of a membrane exists with low dielectric state while the outside with high dielectric state. Parsegian wrote a nice article almost 5 decades ago to demonstrate the dielectric issues (Parsegian, 1969). The membrane generally acts as an insulator with conducting media on both sides.

From simple electrostatic analysis we know the capacitance of a body is defined as the amount of charge needed across it to be stored to create unit potential difference between two terminals. If a potential V can hold a charge Q across a capacitor the capacitance C can be defined as

$$C = Q/V \tag{1.1}$$

A cell membrane structure suggests a model like that where a relatively low dielectric medium is surrounded by two conducting media on both sides (intracellular and extracellular regions). This makes a membrane equivalent to a capacitor.

To calculate the membrane capacitance we need to use the standard electrostatics that is considered under the Coulomb's law, considering an equivalent model structure for membrane which is separation of two parallel conducting plates by an insulating medium. Here membrane (especially the hydrophobic region) is comparable to the insulator medium. The capacitance of a cell membrane can thus be defined as

$$C_m = \kappa\varepsilon_0/d \tag{1.2}$$

K is the dielectric constant for the membrane's inner core and ε_0 is the permittivity of free space. d is the membrane thickness. A low dielectric medium (inner layer) exists between two conducting media (outside membrane). Depending on the values of parameters K, ε_0 and d in various types of cells the values of capacitance of the corresponding membranes vary. But the value is often found to be of the order of 1.0 μF/cm^2.

In model lipid bilayer membrane, under control conditions (Ashrafuzzaman et al., 2006; Hwang et al., 3003; Lundbæk et al., 2005), the membrane capacitance can be measured quite easily using an electrophysiology set up and is found little different. Our measurements of capacitance across dioleoylphosphatidylcholine (DOPC)/n-decane lipid bilayer offered us the value at 3.72 ± 0.24 nF cm^{-2} (mean ± S.D., number of repeats $n = 11$) (see ref. (Ashrafuzzaman et al., 2006)). The capacitance was found so robust in this model lipid bilayer membrane that even due to the effects of amphipathic molecules the value did not change much. We incubated the model DOPC membrane with an anti-fusion peptide Z–Gly–Phe at 800 μM and measured the value at 3.94 ± 0.38 nF cm^{-2} (mean ± S.D., $n = 10$) (Ashrafuzzaman et al., 2006), this value was not much different than the value due the effects of other bilayer-active amphiphiles (Hwang et al., 3003; Lundbæk et al., 2005).

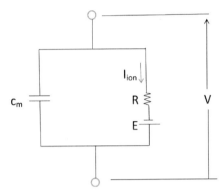

Figure 1.9. Cell membrane based capacitive effects. A parallel capacitor and resistor combination is the schematic form representing the electrical equivalent circuit.

Most importantly, it is mentionable that the cholesterol level, phospholipids and glycolipids, membrane proteins, hydrocarbons, etc. all together are responsible for the certain value of biological cell membrane capacitance. Unlike animal cytoplasmic membrane, bacteria (prokaryotes) does not have cholesterol which may account for a considerable effect on membrane's electrical condition and hence the capacitance.

Understanding of the capacitive effect of membrane helps understand the electrical properties of membrane through a model often referred as Electrical Circuit Model of Cell Membrane (see Figure 1.9).

Here the membrane is assumed to appear with a capacitor in parallel to a resistor. The not necessarily Ohmic resistance acts against the flow of ions across the membrane which is represented by ion current I_{ion}. The capacitive current is $C_m dV/dt$. The capacitive current and the ion current together conserve the current flow between the inside and outside of the membrane. Therefore,

$$C_m dV/dt + I_{ion} = 0 \qquad (1.3)$$

The theoretical calculation of I_{ion} is a long standing challenge. The following Goldman-Hodgkin-Katz (GHK) current equation is one such expression for I_{ion} across membrane:

$$I_{ion} = D/L \, (z^2F^2/RT) \, V \, ([N]_{in} - [N]_{out} \exp\{-zFV/RT\})/(1-\exp\{-zFV/RT\}) \quad (1.4)$$

Here D is the Einstein's diffusion constant, L is the membrane thickness, $[N]_{in}$ and $[N]_{out}$ are ion concentrations inside and outside, respectively, of the cell across the membrane.

1.3. Cell Signaling of Plasma Membranes

Plasma membrane contained lipids, membrane proteins, ion channels, etc. are always engaged in various kinds of signaling processes. The primary role of plasma membrane is to let cell maintain a geometric shape/structure, but among fundamental cellular processes some crucial ones are found to happen across plasma membranes and its vicinity. Bilayer transport of nutrients, various ions, and chemical materials, etc. happen continuously. Slow diffusion, exchange processes and ion channel transport of selective materials are among the mentionable gross techniques that happen nonstop across lipid bilayer membranes.

In maintaining these mechanisms the liquid crystal membrane structure maintains a continuous dynamic state in it where intra and inter layer flipping of lipids, movement of proteins, happen to maintain physiologically required plasma membrane structures. Various mechanisms related to membrane adsorption of materials, interactions among membrane constituents, e.g., among various lipids, among various proteins, inter lipid-protein interactions help plasma membrane to participate in crucial cellular signaling pathways.

As protein and lipid composition of each membrane is unique and their distribution is asymmetric (Pelley, 2007), the related signaling is therefore cell type and/or membrane specific. Although there are a lot of general aspects found common among plasma membranes of all cells. Integral membrane proteins may penetrate the membrane partially or may exist as transmembrane proteins interfacing with both the cytosol and external environment. These proteins interact strongly with the membrane lipids through hydrophobic side chains of amino acids and can only be removed by destroying membrane structure with detergent or membrane active solvent.

They are usually composed of multiple α-helices with hydrophobic side chains; cylindrical arrays form pores for transport of polar molecules. Peripheral membrane proteins are loosely associated with the surface of either side of the membrane; they interact with the membrane through hydrogen bonding or salt-bridging with membrane proteins or lipids and can be removed without disrupting the structure of the membrane. The interaction of integral and peripheral membrane proteins with membrane is highly dependent of membrane lipid types, charge profiles of the lipid and membrane protein structures and the membrane dielectric condition (Ashrafuzzaman and Tuszynski, 2011; 2012). Membrane carbohydrates exist only as extracellular covalent attachments to lipids and proteins (e.g., glycoproteins or glycolipids). Carbohydrate structures are highly variable and may be highly antigenic, thereby contributing to the immune recognition of cells.

Grecco et al., discussed how bidirectional signaling across the plasma membrane is achieved by striking a delicate balance between restriction and propagation of information over different scales of time and space and how underlying dynamic mechanisms give rise to rich, context-dependent signaling responses (Grecco et al., 2011). The living plasma membrane having greater amount of dynamic than static components and its extension inside of the cell as endocytic vesicles constitute a system to receive, integrate, and distribute external and internal signals. The rapid accumulation of knowledge on the dynamics and structure of the plasma membrane has prompted major modifications of the textbook fluid-mosaic model (Nicolson, 2014). It is even proposed that the cooperative action of the hierarchical three-tiered mesoscale (2–300 nm) domains—actin-membrane-skeleton induced compartments (40–300 nm), raft domains (2–20 nm), and dynamic protein complex domains (3–10 nm)—is critical for membrane function and distinguishes the plasma membrane from a classical Singer-Nicolson-type model (Kusumi et al., 2012).

The role of plasma membrane in cell signaling covers verge areas. We just hinted on general aspects here. Specific health and disease issue based various signaling have been found associated to plasma membranes. We wish not to elaborate into those wider aspects due to limited scopes we have in this book.

1.4. Nuclear Membrane

The nuclear membrane is the outer layer of the nucleus which is found in both animal and plant cells. It is also known as the nuclear envelope (NE) which is a double membrane layer. It separates the nuclear contents from its outside materials. As the nucleus contains most of the genetic materials the nuclear envelope is mostly engaged in protecting them from the chemical reactions taking place outside the nucleus. Nuclear membrane controls selective unidirectional or bidirectional permeation of materials, specific proteins, etc. across its double layer.

In a 6 decade old paper Watson raised a few important questions and available evidence based observations and conclusions (Watson, 1955). If the interphase nucleus exerted fundamental control over complex cytoplasmic activities as protein synthesis, this control must be transmitted through the nuclear membrane and that such control must be partly mediated by large molecules which can pass between the nucleus and the cytoplasm (Anderson, 1953). Nuclear envelope hosts nuclear pores what are large complex of proteins that allow small molecules and ions to freely pass, or diffuse, into or out of the nucleus. Nuclear pores also allow necessary proteins to enter the nucleus from the cytoplasm. If the proteins have special sequences that indicate they belong in the nucleus.

Fundamentals of Plasma Nuclear and Mitochondrial Lipid ...   23

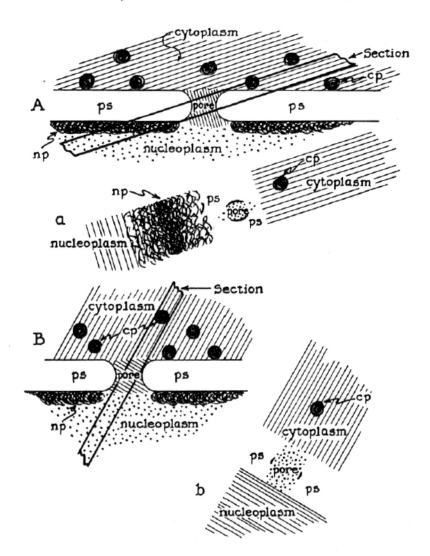

Figure 1.10. Schematic drawings demonstrating the appearance of pores in the nuclear envelope in sections oblique to the surface of the nucleus. In figs. a and b show the appearance of pores to be expected when the section is oriented. In Figure a, the section is nearly tangential to the nuclear surface, thus, the pore appears almost in its entirety and is completely surrounded by the perinuclear space, ps. In Figure b, on the other hand, the section is more nearly normal to the nuclear surface, hence, only a small part of the circumference of the pore appears and a strip of material bridges the gap between the nucleoplasm and the cytoplasm. Symbols: ps ~ perinuclear space; cp ~ cytoplasmic particle; np ~ nuclear particulate matter. Copied from ref. (Watson, 1955) to demonstrate the old day concepts about the nuclear pores and their physiological functions.

Micromanipulation and microdissection of cells by Kite more than a century ago (Kite, 1913) and later by Chambers and Fell (Chambers and Fell, 1931) demonstrated the presence of a membrane around the nuclei of cells in interphase which resisted somewhat the penetration of the dissection needle. Once this barrier was passed the needle could proceed through the nucleus without further interference. Nuclear envelopes teased from cells and examined in the electron microscope were found to consist of two parallel membranes exhibiting, on the surface, numerous ring-shaped structures which were considered to be pores penetrating one or another of the membranes (Callan and Tomlin, 1950; Bairati and Lehmann, 1952; Harris and James, 1952; Gall, 1954). Here we copy Watson's drawings on the Schematic demonstration of the appearance of pores in the nuclear envelope (see Figure 1.10).

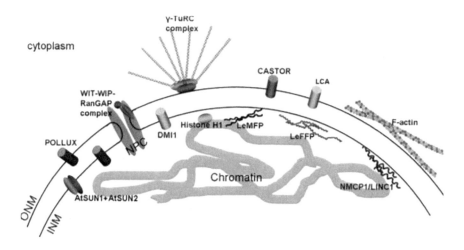

Figure 1.11. Plant NE intrinsic and associated proteins. Membrane-intrinsic plant proteins characterized to date include AtSUN1, AtSUN2, LCA, DMI1, CASTOR, POLLUX as well as WIPs and WITs. Soluble proteins specifically localized at the nuclear periphery include RanGAP, γ-TURC proteins and f-actin on the cytoplasmic face, and NMCP1, LINC1, LeMFP1, LeFFPs and histone H1 at the nucleoplasmic face. It is hypothesized that membrane-intrinsic NE membrane proteins are involved in tethering the soluble proteins to the NE but apart from RanGAP anchoring, the NE proteins and mechanisms involved in this process remain to be elucidated. RanGAP anchorage is mediated via WIP-WIT complexes and also requires the presence of SUN proteins (Zhou et al., 2009; Graumann and Evans, 2010a; Graumann and Evans, 2010).

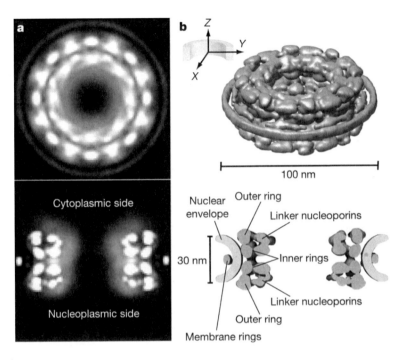

Figure 1.12. a, Projections of nucleoporin mass density (white), derived from the combined localization volumes of all structured domains and the normalized localization probability of all unstructured regions (Alber et al., 2007). Top, *en face* view showing a density projection along the Z-axis from $Z = -50$ nm to $Z = +50$ nm. As in electron microscopy maps of the NPC, radial arms of density correspond to spokes that interconnect to form two strong concentric rings encircling a central region containing low-density unstructured material and bounded by peripheral membrane rings, giving an overall diameter of ~98 nm. Bottom, a slice along the central z-axis showing a projection of density from $X = -5$ nm to $X = +5$ nm. More density can be seen on the cytoplasmic side of the NPC. The low-density unstructured material constricts the central channel to ~10 nm diameter. b, The structured nucleoporin domains of the NPC, represented by a density contour (blue) such that the volume of the contour corresponds approximately to the combined volume of the 456 nucleoporins comprising the NPC. Top: view from a point ~30° from the equatorial plane of the NPC. Bottom: a slice along the central Z-axis between $X = -5$ nm and $X = 5$ nm, in which the nuclear envelope is also shown (in grey). Major features of the NPC are indicated.

The pores were then known as profiles of the ring-shaped structures. The inner and outer nuclear membranes are continuous with one another and enclose the perinuclear space. The pores contain a diffuse, faintly particulate material. It was already concluded by middle of 20th century that pores in the nuclear envelope are a fundamental feature of all resting cells. In certain cells, the outer nuclear membrane is continuous with membranes of the endoplasmic reticulum, hence the perinuclear space is continuous with cavities enclosed by those membranes. There were indications that this was true for all resting cells, at least in a transitory way. On the basis of those observations, a hypothesis was made that two pathways of exchange exist between the nucleus and the cytoplasm; by way of the perinuclear space and cavities of the endoplasmic reticulum and by way of the pores in the nuclear envelope.

Recent investigations based on historical progression have produced fine details about general nuclear membrane structures and specific nuclear membrane pore structures. The nuclear pore complex plays a gatekeeper's role for traffic between the cytoplasm and the nuclear interior region. It is a large supramolecular complex made up of multiple copies of about 30 different proteins — 456 protein molecules in all. Many of the proteins described in animal and fungal systems have not been identified in plants (Graumann and Evans, 2017). Here they have presented a simplified model on the nuclear envelope as presented in Figure 1.11.

A decade ago Alber et al., investigated on molecular architecture of the nuclear pore complex (NPC) (see Figure 1.12) of the yeast cells (Alber et al., 2007). This study revealed that half of the nuclear pore complex is made up of a core scaffold, which is structurally analogous to vesicle-coating complexes. This scaffold forms an interlaced network that coats the entire curved surface of the nuclear envelope membrane within which the complex is embedded. The selective barrier for transport is formed by large numbers of proteins with disordered regions that line the inner face of the scaffold.

1.5. Cell Signaling of Nuclear Membrane

The nuclear envelope is composed of the nuclear membranes, nuclear pore complex, and nuclear lamina. For a detailed structure see Figure 1.13 (Dauer and Worman, 2009) besides Figures 1.11 (nuclear membrane modeling) and 1.12 (nuclear pore complex). The inner and outer nuclear membranes are separated by the perinuclear space, a continuation of the endoplasmic reticulum lumen.

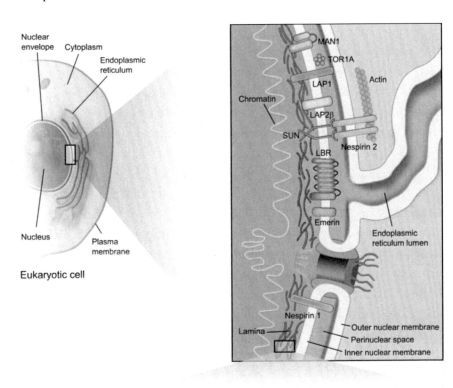

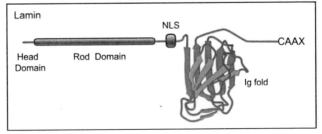

Figure 1.13. Details of nuclear envelope and nuclear lamina.

As a result, proteins secreted into the endoplasmic reticulum, such as torsinA (TOR1A), can potentially reach the perinuclear space; a disease-causing torsinA variant preferentially accumulates in the perinuclear space by binding to lamina-associated protein-1. Some proteins, such as large nesprin-2 isoforms, concentrate in the outer nuclear membrane by binding to SUN proteins within the perinuclear space. The nuclear lamina is a meshwork of intermediate filaments on the inner aspect of the inner nuclear membrane and is composed of proteins called lamins. The lamina is associated with integral proteins of the inner nuclear membrane and representative examples—MAN1, lamina-associated polypeptide-1 (LAP1), the SUN protein lamina-associated polypeptide-2β (LAP2 2β), lamin B receptor (LBR), emerin, and a nesprin-1isoform—are shown. The general structure of a lamin molecule is shown in the lower inset (not to scale). Lamins have α-helical rod domains that are highly conserved among all intermediate filament proteins and are critical for the formation of dimers and higher-ordered filaments. They have head and tail domains that vary in sequence among members of intermediate filament protein family. Within the tail domain, lamins contain a nuclear localization signal (NLS) and an immunoglobulin-like fold (Ig fold). Most lamins (not mammalian lamin C or C2) contain at their carboxyl-termini a CAAX motif that acts as a signal to trigger a series of chemical reactions leading to protein modification by fanesylation and carboxymethylation.

There are a lot of studies recently made to uncover new functions of nuclear envelope proteins and underlie an emerging view of the nuclear envelope as a critical signaling node in development and disease (see a bulk of references mentioned in the article (Dauer and Worman, 2009)). The first indication of a role for the NE in calcium signaling was elucidated through calcium pumping ATPase of the sarcoplasmic reticulum/endoplasmic reticulum (SERCA) type called LCA (Downie et al., 1998). The identification of a NE Ca^{2+} signaling pathway involved in mycorrhizal infection and nodulation with the nuclear periplasm acting as a Ca^{2+} signalling pool (Chabaud et al., 2011) Arbuscular mycorrhizal hyphopodia and germinated spore exudates trigger Ca^{2+} spiking in the legume and nonlegume root epidermis. *New Phytologist* 189, 347–355]. Significant advances have been made through identifying the plant nucleoporins (Tamura et al., 2010).

1.6. MITOCHONDRIAL MEMBRANE

Mitochondria are separated from the cytoplasm by two membranes, namely, the outer and inner mitochondrial membranes (see Figure 1.14). The outer and inner mitochondrial membranes are geometrically and electrically distinguishable. The inner mitochondrial membrane area is larger than that of outer membrane. For typical liver mitochondria, the inner membrane area is about 5 times as large as that of outer membrane due to cristae. The outer membrane is porous while the inner membrane is a tight barrier. Outer membrane has no membrane potential while the inner membrane has (for details one may read ref. (Zorova et al., 2018)). The trans inner mitochondrial membrane potential is about 180 mV with negative inside.

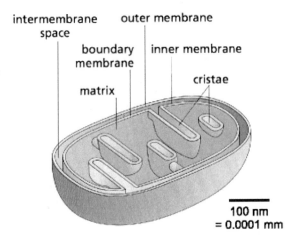

Figure 1.14. Membrane compartments in the mitochondrion. The outer membrane separates mitochondria from the cytoplasm. It surrounds the inner membrane, which separates the inter-membrane space from the protein-dense central matrix. The inner membrane is differentiated into the inner boundary membrane and the cristae. The two regions are continuous at the cristae junctions. The cristae extend more or less deeply into the matrix and are the main sites of mitochondrial energy conversion. The shallow proton gradient between the inter-membrane space (pH 7.2–7.4) and the matrix (pH 7.9–8) drives ATP production by the ATP synthase in the membranes of the cristae. (Adapted from Figure 14–8 C in (Alberts et al., 2014) with permission of the publisher). Copied from ref. (Kühlbrandt, 2015); Adapted from Figure 14–8 C in (Alberts et al., 2014) with permission of the publisher.

Ions and small, uncharged molecules freely move through porous outer membrane while the inner membrane acts as diffusion barrier to ions and molecules. Larger molecules, especially proteins, have to be imported across outer membranes by special translocases. The oxidative phosphorylation takes place in an inner membrane suite of membrane protein complexes that create the electrochemical gradient across the inner membrane, or use it for ATP synthesis. The mitochondrial membrane potential ($\Delta\Psi_m$) generated by proton pumps is an essential component in the process of energy storage during oxidative phosphorylation (see Figure 1.15). The mitochondrial membrane potential raised driving force prefers for inward transport of cations and outward transport of anions. This property thus allows accumulation of metal cations in the mitochondria exerted by intrinsic *electrogenic* transporters and depending on inner mitochondrial membrane potential.

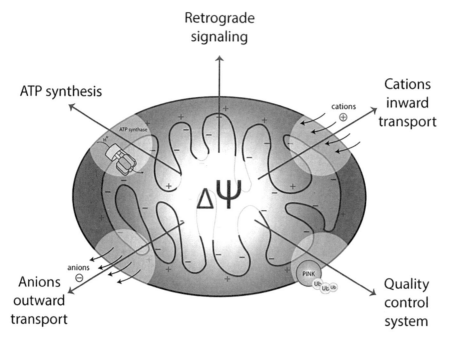

Figure 1.15. Mitochondrial membrane transport and related control systems. Copied from ref. (Zorova et al., 2018).

1.7. MITOCHONDRIAL MEMBRANE CONSTITUENTS AND THEIR ROLES IN SIGNALING

Mitochondrial membrane constituents continuously participate in signaling processes. Most of these signaling are related to cell health and diseases. Various lipids and proteins are mostly involved. We shall briefly address them here.

1.7.1. Mitochondrial Membrane Lipids and Their Roles in Signaling

The mitochondrial inner membrane and outer membrane contain varied compositions of phospholipids, cholesterol, and proteins. Majority of the structural, functional, and signaling aspects of mitochondrial membranes and the intermembrane space rely on the organization, dynamics, mutual interactions, and domain formation of various types of lipids in membranes. Biological energy conversion is carried out by the membrane protein complexes of the respiratory chain and the mitochondrial ATP synthase in the inner mitochondrial membrane cristae. Lipids participate strongly in formation of membrane transport events like channels or pores, so the lipid specific regulation of channel transport mechanisms is a crucial aspect related to membrane functions. Major membrane lipid species are found to play important roles in cell signaling associated specifically with apoptosis. Lipid organization and function may vary between normal and pathological cellular conditions. Specialized lipid specific mitochondrial membrane transport may regulate the cellular health condition, and malfunctioning of such transport mechanisms may play critical roles in raising cell based disease states.

Tables 1.1 and 1.2 present the lipid composition in mitochondria and mitochondrial outer membrane vesicles (copied from ref. (Anton et al., 1997)). Analysis of the phospholipid composition of the OMV showed that phosphatidylcholine, phosphatidylethanolamine and phosphatidylinositol are the major phospholipid constituents, and that cardiolipin is only present in trace amounts. The phospholipid composition is very similar to that of the highly purified OMV from mitochondria of *Neurospora crassa*, although the latter still contain a small amount of cardiolipin. Cardiolipin is localized mostly in the inner mitochondrial membrane (Lewis and McElhaney, 2009). The role of cardiolipin in mitochondrial membrane organization is quite rigorously addressed in ref. (Brown et al., 2013). Here importantly discussed on the issue that many of the proteins involved in mitochondrial transport processes depend on the presence of cardiolipin micro-domains (see Figure 1.16) in the inner mitochondrial membrane. The inner mitochondrial membrane is somewhat similar in lipid composition to the bacterial membrane. Inspecting the pig heart mitochondria, the phospholipid composition is found as follows: phosphatidylethanolamine 37.0%, Phosphatidylcholine 26.5%, cardiolipin 25.4%, and phosphatidylinositol 4.5% (Comte et al., 1976).

Table 1.1. Phospholipid composition

	Mitochondria	OMV
Phospholipid/protein	220 ± 20	1220 ± 150
Phosphatidylcholine	44.3 ± 2.2	54.7 ± 3.1
Phosphatidylethanolamine	34.4 ± 2.0	27.5 ± 1.7
Cardiolipin	13.5 ± 2.1	0.3 ± 0.3
Phosphatidylinositol	5.4 ± 0.8	13.4 ± 2.3
Phosphatidylserine	0.5 ± 0.2	2.1 ± 0.5
Phosphatidic acid	nd	0.4 ± 0.5
Phosphatidylglycerol	0.1 ± 0.1	0.2 ± 0.3
Lyso-phospholipids	0.4 ± 0.4	0.4 ± 0.6
Sphingomyelin	0.9 ± 0.4	0.7 ± 0.6

* Phospholipid composition of rat liver mitochondria and mitochondrial outer membrane (OMV) vesicles prepared by the swell–disruption (SD) method. The numbers represent phospholipid phosphorus expressed as percentages of total lipid phosphorus, and are presented as mean values ± S.D., which derive from six batches of mitochondria and 11 batches of OMV; the phospholipid-to-protein ratios are expressed as nmol of phospholipid phosphorus per mg of protein. nd, not detectable

In S. cerevisiae mitochondrial inner membrane, phosphatidylcholine 38.4%, phosphatidylethanolamine 24.0%, phosphate-idylinositol 16.2%, cardiolipin 16.1%, phosphatidylserine 3.8%, and phosphatidic acid 1.5% (Lomize et al., 2013).

Table 1.2. Phospholipid composition of mitochondria and mitochondrial outer membrane vesicles from *Neurospora crassa*.

	Mitochondria	OMV
Phospholipid/protein	490 ± 50	2580 ± 240
Phosphatidylcholine	36.7 ± 3.0	50.2 ± 1.6
Phosphatidylethanolamine	39.9 ± 1.0	32.7 ± 2.6
Cardiolipin	11.8 ± 0.9	3.1 ± 0.5
Phosphatidylinositol	4.5 ± 0.3	9.2 ± 1.1
Phosphatidylserine	0.7 ± 0.2	0.6 ± 0.2
Phosphatidic acid	0.6 ± 0.2	0.7 ± 0.4
Phosphatidylglycerol	1.0 ± 0.4	0.4 ± 0.3
Lyso-phospholipids	2.3 ± 1.7	1.2 ± 1.1
PX[a]	1.5 ± 0.3	1.7 ± 0.3

[a]PX denotes an unidentified phosphorus-containing lipid.

The numbers represent phospholipid phosphorus expressed as percentages of total lipid phosphorus, and are presented as mean values ± S.D., which derive from four batches of mitochondria and six batches of OMV. The phospholipid-to-protein ratios are expressed as nmol of phospholipid phosphorus per mg of protein

Ceramide, a membrane sphingolipid, participates in forming channels in the mitochondrial outer membrane and plays a major regulatory role in apoptosis by inducing the release of proapoptotic proteins from the mitochondria (Siskind, 2005). Mitochondria contain the enzymes responsible for the synthesis and hydrolysis of ceramide, there exists a mechanism for regulating the level of ceramide in mitochondria. Ceramide, is generated by de novo synthesis pathway in the endoplasmic reticulum (ER).Ceramide accumulation in mitochondria has been associated with reperfusion damage mainly due to the observation that the addition of this sphingolipid to isolated mitochondria, mimetizes some of the events observed in these organelles after increased outer membrane permeability (Siskind et al., 2006).

Mitochondrial outer and inner membrane based protein-lipid relative compositions are different. In the mitochondrial inner membrane, the protein-to-lipid ratio is 80:20, while in mitochondrial outer membrane the ratio is 50:50 (Stefan, 2001).

In Table 1.1 the Percoll gradient-purified mitochondria from rat liver and derived OMV were analyzed for their phospholipid content. In Table 1.2 the phospholipid compositions of mitochondria and highly purified OMV from *Neurospora crassa* are compared. The phospholipid composition of the mitochondria resembles that of rat liver mitochondria (Table 1.1) with PE constituting the major class of phospholipids.

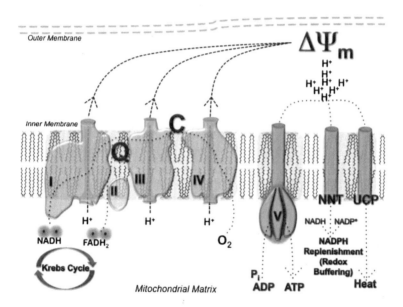

Figure 1.16. The proton circuit in healthy bioenergetics. Respiratory complexes are tightly arranged in functional "supercomplexes" or "respirasomes." Cardiolipin-enriched microdomains, depicted in light blue, are necessary for optimal protein organization and function. The proton-motive force consists primarily of the mitochondrial membrane potential, $\Delta\psi_m$. The potential energy in $\Delta\psi_m$ can be utilized to regenerate ATP, replenish cellular redox pools through the nicotinamide nucleotide transhydrogenase (NNT), generate heat through uncoupling proteins (UCP), or promote trans-membrane transport (not depicted). Copied from ref. (Brown et al., 2013).

1.7.2. Mitochondrial Proteins and Membrane Based Signaling

Mitochondria have versatile functional proteins, many of whom are connection to cell health condition and rise of diseases (for a detail summary see (Ashrafuzzaman, 2018). As discussed earlier mitochondria are subdivided into several compartments: the outer and inner membranes, and the matrix space (see Figure 1.17). Despite the presence of 13 proteins that are mitochondrially encoded and part of the bioenergetic oxidative phosphorylation (OXPHOS) system, the complete repertoire of mitochondrial proteins (including further subunits of the OXPHOS) has to be imported via the translocase of the outer membrane (TOM) (Straub et al., 2016) and the translocase of the inner membrane (TIM) (Mokranjac and Neupert, 2010).

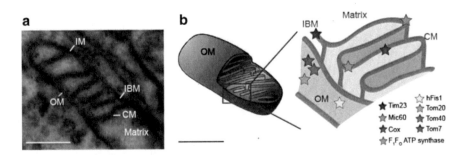

Figure 1.17. Ultrastructure of mitochondria and respective localization of important proteins. a Electron microscopy image showing part of a mitochondrion, with the membranes appearing black: OM outer membrane, IM inner membrane [which is partitioned into the inner boundary membrane (IBM) and the cristae membranes (CM)]. The matrix is an aqueous compartment. b Scheme showing the localization of the proteins that have been studied in terms of mobility and localization. Tom40 Core subunit of the translocase of the outer membrane (TOM) complex, Tom20 receptor subunit of TOM, Tom7 subunit of TOM, hFis human fission factor in the OM, Mic60 mitofilin, part of the MICOS complex in the IBM, Tim23 part of the inner membrane translocase (TIM) in the IBM, Cox cytochrome c oxidase, a complex of the oxidative phosphorylation (OXPHOS) system in the CM, F 1 F 0 ATP synthase, a further OXPHOS complex. Copied from ref. (Appelhans and Busch, 2017).

Furthermore, the complex IM architecture is under the control of several proteins, including F1FOATP synthase (Gavin et al., 2004) and mitofilin/Mic60 (John et al., 2005; Tarasenko et al., 2017). Mitochondrial membranes are involved in various transport mechanisms that release proteins and other materials that are directly linked to cellular functions including apoptosis, rise of diseases, etc. We have taken this issue quite in details in my recent book (Ashrafuzzaman, 2018). This is why I wish not to elaborate it here. Instead the readers are encouraged to go through my book (see chapter 8, ref. (Ashrafuzzaman, 2018)) for details.

REFERENCES

Alberts B, Johnson A, Lewis J, et al., 2002. *Molecular Biology of the Cell*. 4th edition. New York: Garland Science.

Alberts B., Johnson A., Lewis J., Morgan D., Raff M., Roberts K., Walter P. 2014. *Molecular Biology of the Cell*, Sixth Edition. W. W. Norton & Company (UK).

Alber F., Dokudovskaya S., Veenhoff L. M., Zhang W., Kipper J., Devos D., Suprapto A., Karni-Schmidt O., Williams R., Chait B. T., Rout M. P., Sali A. 2007. Determining the architectures of macromolecular assemblies. *Nature* 450: 683–694.

Anderson N. G. 1953. On the Nuclear Envelope. *Science* 117: 517-521.

Anton I. P., de Kroon M., Mayer A., Kruijff B. 1997. Phospholipid composition of highly purified mitochondrial outer membranes of rat liver and Neurospora crassa. Is cardiolipin present in the mitochondrial outer membrane?, *Biochimica et Biophysica Acta (BBA) – Biomembranes* 1325: 108-116.

Appelhans T. and Busch K. B. 2017. Dynamic imaging of mitochondrial membrane proteins in specific sub-organelle membrane locations. *Biophys Rev.* 9(4): 345–352.

Ashrafuzzaman, M. 2018. *Nanoscale Biophysics*, Springer International Publisher, Springer Nature Switzerland AG.

Ashrafuzzaman M., Tuszynski J. 2012. Regulation of Channel Function Due to Coupling with a Lipid Bilayer. *J. Comp. Theor. Nanosci.* 9: 564-570.

Ashrafuzzaman M. and Tuszynski J. 2013. *Membrane Biophysics*. Springer-Verlag Berlin Heidelberg.

Ashrafuzzaman M., Lampson M. A., Greathouse D. V., Koeppe R. E., II, Andersen O. S. 2006. Manipulating lipid bilayer material properties using biologically active amphipathic molecules. *J. Phys.: Condens. Matter* 18, S1235–S1255.

Bairati A., and Lehmann F. 1952. *Experienlia* 8: 60.

Brown D. A., Hani N. Sabbah, Saame RazaShaikh, Mitochondrial inner membrane lipids and proteins as targets for decreasing cardiac ischemia/reperfusion injury, *Pharmacology & Therapeutics*, Volume 140, Issue 3, December 2013, Pages 258-266.

Brown D. A., Sabbah H. N., Shaikh S. R. 2013. Mitochondrial inner membrane lipids and proteins as targets for decreasing cardiac ischemia/reperfusion injury, *Pharmacology & Therapeutics* 140: 258-266.

Callan H. G., and Tomlin S. G. 1950. *Proc. Roy. Soc.* London, Series B 137: 367.

Chabaud M., Genre A., Sieberer B. J. et al., 2011. Arbuscular mycorrhizal hyphopodia and germinated spore exudates trigger Ca2+ spiking in the legume and nonlegume root epidermis. *New Phytologist* 189: 347–355.

Chambers R., and Fell B. 1931. *Proc. Roy. Soc.* London, Ser/es B 109: 380.

Comte J., Maïsterrena B., Gautheron D. C. 1976. "Lipid composition and protein profiles of outer and inner membranes from pig heart mitochondria. Comparison with microsomes." *Biochim. Biophys.* Acta. 419: 271–84.

Dauer W. T., Worman H. J. 2009. The Nuclear Envelope as a Signaling Node in Development and Disease. *Developmental Cell* 17: 626-63.

Deamer D. 2017. The Role of Lipid Membranes in Life's Origin. *Life* (Basel). 7(1): 5.

Dowhan W., M. Bogdanov. 2002. Functional roles of lipids in membranes. *New Comprehensive Biochemistry*, Volume 36: Pages 1-35 (ch. 1).

Dowhan W, M. Bogdanov. 2002. In: *Biochemistry of Lipids, Lipoproteins and Membranes*. Vance DE, Vance JE, editors. Vol. 36. Elsevier; Amsterdam: pp. 1–35.

Downie L., Priddle J., Hawes C., and Evans D. E. 1998. A calcium pump at the higher plant nuclear envelope. *Febs Letters* 429: 44–48.

Escribá P. V., Gonzáles-Ros, J. M., F. M. Goñi, P. K. J. Kinnunen, L. Vigh, L. Sánchez-Magraner, A. M. Fernández, X. Busquets, I. Horváth, G. Barceló-Coblijn. 2008. Membranes: a meeting point for lipids, proteins and therapies. *J. Cell. Mol. Med.* 12: 829-875.

Frank A., Svetlana D., Veenhoff L. M., Zhang W., Kipper J., Devos D., Suprapto A., Karni-Schmidt O., Williams R., Chait B. T., Sali A., Rout M. P. 2007. The molecular architecture of the nuclear pore complex, *Nature* 450: 695–701.

Gall J. G. 1954. *Exp. Cell Research* 7: 197.

Gavin P. D., Prescott M., Luff S. E., Devenish R. J. 2004. Cross-linking ATP synthase complexes in vivo eliminates mitochondrial cristae. *J Cell Sci.* 117(Pt 11):2333-43.

Graumann K., Evans D. E., 2017. *The Nuclear Envelope – Structure and Protein Interactions, Annual Plant Reviews book series*, Volume 46: Plant Nuclear Structure, Genome Architecture and Gene Regulation, https://doi.org/10.1002/9781119312994.apr0498.

Graumann K. and Evans D. E. 2010. *The plant nuclear envelope in focus*. Biochemical Society Transactions 38, 307–311 © Portland Press Limited and International Federation for Cell Biology.

Grecco H. E., Schmick M., and Bastiaens P. I. H. 2011. Signaling from the Living Plasma Membrane. *Cell* 144: 897-909.

Harris P., and James T. 1952. *Gxperientia* 8: 384.

Huang H. W. 1986. Deformation free energy of bilayer membrane and its effect on gramicidin channel lifetime. *Biophys. J.* 50, 1061–1070.

Hwang T. C., Koeppe R. E., II, Andersen O. S. 2003. Genistein can modulate channel function by a phosphorylation-independent mechanism: importance of hydrophobic mismatch and bilayer mechanics. *Biochemistry* 42, 13 646–13 658.

Jacobson K., E. D. Sheets, R Simson. 1995. Revisiting the fluid mosaic model of membranes. *Science* 268: 1441-1442.

Jain SK. 1988. Evidence for membrane lipid peroxidation during the in vivo aging of human erythrocytes. *Biochim Biophys Acta.* 937(2):205-10.

John GB, Shang Y, Li L, Renken C, Mannella CA, Selker JM, Rangell L, Bennett MJ, Zha J. 2005. The mitochondrial inner membrane protein mitofilin controls cristae morphology. *Mol Biol Cell.* 16(3):1543-54.

Kite G. L. 1913. *Am. jr. Physiol.* 32: 146.

Kucerka N., S. Tristram-Nagle, J. F. Nagle. 2006. Closer look at structure of fully hydrated fluid phase DPPC bilayers. *Biophys J.* 90: L83–L85.

Kühlbrandt W. 2015. Structure and function of mitochondrial membrane protein complexes. *BMC Biology* 13:89, https://doi.org/10.1186/s 12915-015-0201-x.

Kusumi A., Fujiwara T. K., Chadda R., Xie M., Tsunoyama T. A., Kalay Z., Kasai R. S., and Suzuki K. G. N. Dynamic Organizing Principles Of The Plasma Membrane That Regulate Signal Transduction: Commemorating The Fortieth Anniversary Of Singer And Nicolson's Fluid-Mosaic Model. 2012. An. Rev. *Cell and Developmental Biology* 28: 215-250.

Leah J. Siskind (2005) Mitochondrial Ceramide and the Induction of Apoptosis. *J Bioenerg Biomembr* 37: 143–153.

Lewis R. N. A. H., R. N. McElhaney. 2009. The physicochemical properties of cardiolipin bilayers and cardiolipin-containing lipid membranes. *Biochim Biophys Acta Biomembr*, 1788: 2069-2079.

Lin S., Huang, T. T., Shen B. 2012. Chapter Sixteen - Tailoring Enzymes Acting on Carrier Protein-Tethered Substrates in *Natural Product Biosynthesis, Methods in Enzymology* 516: 321-343.

Lomize A., Lomize M., Pogozheva I. 2013. "*Membrane Protein Lipid Composition Atlas.*" Orientations of Proteins in Membranes. University of Michigan. Retrieved 10 April 2014.

Lundbæk J. A., Birn P., Tape S. E., Toombes G. E., Søgaard R., Koeppe R. E., II, Gruner S. M., Hansen A. J., Andersen O. S. 2005. Capsaicin regulates voltage-dependent sodium channels by altering lipid bilayer elasticity. *Mol. Pharmacol.* 68, 680–689.

Marsh D. 2007. Lateral pressure profile, spontaneous curvature frustration, and the incorporation and conformation of proteins in membranes. *Biophys J.* 93: 3884–3899.

Mokranjac D., Neupert W. 2010. The many faces of the mitochondrial TIM23 complex. *Biochim Biophys Acta.* 1797(6-7):1045-54.

Nicolson G. L. 1976. Transmembrane control of the receptors on normal and tumor cells. I. Cytoplasmic influence over cell surface components. *Biochim. Biophys. Acta*, 457: 57-108.

Nicolson G. L. 2014. The Fluid—Mosaic Model of Membrane Structure: Still relevant to understanding the structure, function and dynamics of biological membranes after more than 40 years. *Biochimica et Biophysica Acta (BBA) – Biomembranes.* 1838: 1451-1466.

Nicolson G. L., T. Ji, G. Poste. 1977. *The dynamics of cell membrane organization.* Poste G., G. L. Nicolson (Eds.), Dynamic Aspects of Cell Surface Organization, Elsevier, New York: 1-73.

Parsegian A. 1969. Energy of an Ion crossing a Low Dielectric Membrane: Solutions to Four Relevant Electrostatic Problems. *Nature* 221, 844–846.

Pelley J. W. 2007. *Membranes and Intracellular Signal Transduction.* Elsevier's Integrated Biochemistry, Pages 37-46.

Singer S. J., Nicolson G. L. 1972. The fluid mosaic model of the structure of cell membranes. *Science.* 175(4023): 720-31.

Siskind L. J., R. N. Kolesnick, M. Colombini. 2006. Ceramide forms channels in mitochondrial outer membranes at physiologically relevant concentrations. *Mitochondrion*, 6, pp. 118-125.

Sodt, A. J., Venable R. M., Lyman E., Pastor RW. 2016. Non-additive compositional curvature energetics of lipid bilayers. *Phys. Rev. Lett.* 117, 138104.

Stefan K. (2001). "*Mitochondria: Structure and Role in Respiration*" (PDF). Nature Publishing Group. Archived from the original (PDF) on 21 October 2012. Retrieved 9 April 2014.

Straub S. P., Stiller S. B., Wiedemann N., Pfanner N. 2016. Dynamic organization of the mitochondrial protein import machinery. *Biol Chem.* 397(11):1097-1114.

Tamura K., Fukao Y., Iwamoto M., Haraguchi T., and Hara-Nishimura I. 2010. Identification and characterization of nuclear pore complex components in Arabidopsis thaliana. *Plant Cell* 22: 4084–4097.

Tarasenko D, Barbot M, Jans DC, Kroppen B, Sadowski B, Heim G, Möbius W, Jakobs S, Meinecke M. 2017. The MICOS component Mic60 displays a conserved membrane-bending activity that is necessary for normal cristae morphology. *J Cell Biol.* 216(4):889-899.

van Meer G, Lisman Q. 2002. Sphingolipid transport: rafts and translocators. *J Biol Chem.* 277(29):25855-8.

van Meer G., Voelker D. R. and Feigenson G. W. 2008. Membrane lipids: where they are and how they behave. *Nat Rev Mol Cell Biol.* 9(2): 112–124.

Watson M. L. 1955. The nuclear envelope. *J. Biophysic. And Biochem. Cytol.*, Vol. I, No. 3.

Zhou K., Rolls M. M., Hall D. H., Malone C. J., and Hanna-Rose W. 2009. A Zyg-12-dynein interaction at the nuclear envelope defines cytoskeletal architecture in the C. elegand gonad. *Journal of Cell Biology* 186: 229–241.

Zorova L. D., Popkov V. A., Plotnikov E. Y., Silachev D. N., Pevzner I. B., Jankauskas S. S., Babenko V. A., Zorov S. D., Balakireva A. V., Juhaszova M., Sollott S. J., Zorov D. B. 2018. Mitochondrial membrane potential. *Analytical Biochemistry* 552: 50-59.

In: Lipid Bilayers
Editor: Mohammad Ashrafuzzaman
ISBN: 978-1-53616-392-6
© 2019 Nova Science Publishers, Inc.

Chapter 2

MECHANICAL STRESSES IN THE LIPID BILAYER OF ERYTHROCYTE MEMBRANES

Pavel V. Mokrushnikov[*], *PhD*
Department of Physics,
Novosibirsk State University of Architecture
and Civil Engineering, (SIBSTRIN), Novosibirsk, Russia

ABSTRACT

The effect of mechanical stresses in the membrane on the erythrocyte shape and surface morphology was studied. A normal erythrocyte has the shape of a biconcave discocyte with alternating convex and concave regions of the membrane. There are no direct experimental data on mechanical stresses in these regions. Atomic force microscopy of the erythrocyte membrane surface and IR spectroscopy of membranes were used to reveal the distribution of mechanical stresses in biomembrane. A situation when the membrane stability is impaired and folds appear on its surface is considered. The increasing mechanical stresses in the membrane can result in its rupture. The mechanism of such a rupture is discussed. It

---

[*] Corresponding Author's Email: pwm64@ngs.ru.

is concluded that mechanical stresses in the cell cytoskeleton prevent rupture of the plasmatic membrane.

1. INTRODUCTION

Mechanical stress is among the types of stimulation perceived by a living cell. The degree and implications of mechanical deformations and vibrations for cell activity will depend on intrinsic mechanical characteristics of the cell and sensitivity of its mechanosensors. Mechanosensors can be represented by various structures that are able to perceive mechanical deformations and vibrations and respond to them [1]. The first candidate for the role of mechanosensor is the extracellular matrix with the bound membrane proteins. It was shown that the tensile force applied to a culture of neurons or smooth muscle cells through the extracellular matrix enhances polymerization of microtubules [2, 3].

On the cell membrane surface, various proteins can form a focal-adhesive complex in the zones of integrin contacts with adhesive proteins of the extracellular matrix. It was supposed that an external mechanical force can produce conformational changes in one or several proteins of the focal-adhesive complex, thus triggering a cascade of subjacent signaling pathways [4]. The inverse supposition can also be made – changes in the conformation of membrane proteins can alter mechanical stresses in biomembrane.

Deformation of the membrane affects the cation transport activity of mechanosensitive channels as a result of conformational changes in the lipid bilayer or portal domains of the channel itself [4, 5, 6]. In eukaryotic cells, mechanosensitive channels are represented by epithelial sodium channels ENaCs – a family of ion channels from the superfamily of degenerin/ENaC (DEG/ENaC), which were detected in the cells of different sodium absorbing types of epithelium [7].

Mechanical stresses in biomembranes correlate with biomembrane microviscosity. The higher is the mechanical compressive stresses in biomembrane, the greater is its microviscosity. Microviscosity of erythrocyte membranes affects the activity of enzymes, for example, carboanhydrase II. In experiments, erythrocyte membranes were donated from patients with diabetes and hypercholesterolemia and from healthy persons. Membranes from patients had an increased microviscosity as compared to healthy persons. The activity of carboanhydrase II was measured electrometrically. An increased activity of this enzyme was found in the membranes of patients. In patients with hypercholesterolemia, the activity of carboanhydrase II showed a more pronounced increase in comparison with diabetes patients [8].

The submembrane cytoskeleton is also involved in regulation of the ion channel activity. Cytochalasin D binds to the rapidly growing end of the actin filament and blocks (sometimes incompletely) both the attachment and detachment of subunits at this end. In the experiment, treatment of the cell culture K562 with cytochalasin D activated sodium channels, whereas polymerization of actin on cytoplasmatic side of the cell membrane led to their inactivation [9]. Therewith, in the cells of K562 line, fragmentation of actin filaments associated with plasmatic membrane can be the main factor determining the activity of sodium channels in response to increased intracellular concentration of calcium ions [10].

A study performed by the patch-clamp method revealed that actin microfilaments participate in the regulation of chlorine channels [11, 12], Na^+,K^+-ATPase [13], electrically excitable sodium channels in brain cells [14], and sodium channels in reabsorbing epithelial cells [15]. The causes and mechanism of this effect are not clear.

When a force is applied through the membrane-bound receptors, this sometimes deforms the cell nucleus [16]; this suggests a direct effect of external forces on chromatin and thus on the level of gene expression [17]. In this case, forces can be transduced through the cytoskeleton network to the nuclear envelope, and then through the laminin network to chromatin. In addition, the action of external forces can be transferred to microtubules, leading to their rupture, depolymerization and triggering of signaling pathways [18].

Catecholamines (adrenaline and noradrenaline) can act on the target cells through adrenoreceptors located on plasmatic membrane. Supposedly, adrenoreceptors reside also on plasmatic membranes of erythrocytes and are connected with the spectrin-ankyrin-actin network [19]. Steroid hormones (cortisol, testosterone, androsterone) can also bind to adrenoreceptors in plasmatic membrane [20]. Binding of hormones with the cell receptors plays an important role in the regulation of gene expression [21]. The interaction of hormones and cells triggers a cascade of biochemical reactions. Stress hormones (cortisol, adrenaline, noradrenaline) are involved in the enhancement of energy exchange in the organism. This is their main role in implementation of different adaptation mechanisms under extreme conditions. This function of hormones is well understood now. However, the indicated hormones exert also the non-genomic effect on cells [20].

Stress hormones (for example, cortisol and adrenaline) and androgens (such as androsterone, testosterone, DHEA, and DHEAS) can bind to plasmatic membranes, change their structure and increase their microviscosity, thus affecting the functions of membrane and cell (non-genomic effect) [22, 23, 24]. The structure of membranes depends also on blood proteins, for example, apolipoprotein A-1 [25]. Changes in the membrane structure affects the passage of erythrocytes in microcapillaries and the activity of Na^+, K^+- ATPases in erythrocyte membranes [26]. It was shown experimentally that a long-term action of a stress hormone adrenaline can result in slugging of erythrocytes and occlusion of cardiac vessels (stasis) [27]. Measurements of membrane microviscosity under the joint action of adrenaline and cortisol demonstrated that adrenaline is displaced by cortisol from the binding sites on plasmatic membrane [28].

New nanoparticles are being devised nowadays, and their minimum technological limit has reached 1 nm (for example, fullerenes). At present, the annual output of various nanoparticles exceeds hundreds of tons. They are widely employed for surface nanostructuring of machines and mechanisms and are used in cosmetics, entering the composition of creams and scrubs. However, their action on the human organism was investigated insufficiently. Penetrating into the organism, nanocrystals are adsorbed on the cell surface and change the field of mechanical stresses in plasmatic membrane and its functions. If the size of nanoparticles is below the critical value, they can penetrate into the membrane and destroy it. This is why some newfangled cosmetics are harmful substances. For example, the introduction of titania nanocrystals with the sizes below the critical value into scrubs may cause a lysis of healthy epidermal cells [29, 30].

This chapter considers the effect of noradrenaline on the surface morphology of erythrocyte membranes, microviscosity of their lipid bilayer, and activity of Na^+, K^+-ATPases in erythrocyte membranes. The mechanism underlying the effect of hormones on the structural transformations in biomembranes is described. A hypothesis explaining the influence of membrane microviscosity (i.e., mechanical stresses in biomembrane) on the activity of membrane Na^+, K^+-ATPases is proposed. Theoretical models of mechanical stresses in plasmatic membranes are discussed.

2. Mechanical Stresses in the Lipid Bilayer of Erythrocyte Membranes

2.1. The Effect of Noradrenaline on the Surface Morphology of Erythrocyte Membranes

Let us consider the effect of noradrenaline on the structure of erythrocyte membranes and activity of Na^+, K^+-ATPases in erythrocyte membranes. It will be demonstrated that this effect is exerted via changing the structure of the lipid bilayer and mechanical stresses in biomembrane.

Atomic force microscopy of erythrocyte membranes was carried out by the following method. Erythrocytes were isolated from fresh blood taken after decapitation of male Wistar rats under light Nembutal anesthesia. The experiments were carried out in compliance with the Declaration of Helsinki developed by the World Medical Association in 1964.

The obtained blood was diluted twofold with the isotonic phosphate buffer (pH 7.4) containing 0.043 M KH_2PO_4 and 0.136 M Na_2HPO_4. After sedimentation of the cells by centrifugation at 330 g for 10 min, supernatant was decanted and the procedure was repeated twice more.

All the procedures were performed at 4°C. Upon incubation with noradrenaline for 2 min, 20 µL of the resulting erythrocyte suspension was smeared on a microscope slide as a thin layer; after preliminary drying in air, the specimen was examined in a Solver Bio (NT–MDT, Russia) atomic force microscope at a temperature of 24°C under semicontact mode. The NSG11 (NT–MDT, Russia) silicon cantilevers with a resonant frequency from 120 to 180 kHz and stiffness constant ~ 6 H/m (all data provided by the manufacturer) were used. This allowed obtaining images of the erythrocyte membrane relief after the adsorption of noradrenaline molecules. A noradrenaline solution with the concentration of 10^{-6} M was prepared using the isotonic phosphate buffer.

In this work, the specific concentration of noradrenaline was used. According to our measurements, erythrocyte membrane proteins in erythrocyte suspension have a concentration of 0.5 mg protein/mL. If the molar concentration of noradrenaline is divided by the concentration of membrane proteins, the specific concentration of noradrenaline will be obtained. The measurements were performed by Boris Nikolaevich Zaitsev, Senior Researcher at the State Research Center of Virology and Biotechnology VECTOR, Koltsovo, Novosibirsk region, Russia.

The atomic force microscopy examination revealed that erythrocytes of healthy rats look as large biconcave discs (discocytes) with the diameter of ~6 nm. At a greater magnification, a slight folding (the horizontal dimensions 100×100 nm^2 and height 4 – 6 nm) was observed on the surface of these discocytes; this could be caused by the presence of membrane-bound proteins (Figure 1). Such proteins together with the adjacent lipids form the protein-lipid domains that determine the surface morphology of erythrocyte membrane and its functions.

Noradrenaline with the specific concentration of $2 \cdot 10^{-9}$ mol/mg protein produced changes in the surface morphology of erythrocyte membranes (Figure 2). Convex domains with a quasi-chessboard distribution appeared on the membrane surface; they alternated with the regions of considerable surface hollows. Noradrenaline caused the appearance of domains with the size of 100×100 nm^2 and height 2–3 nm.

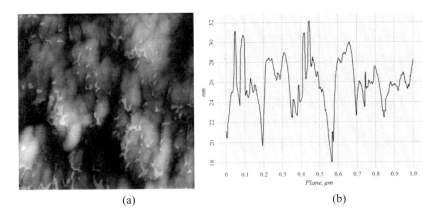

Figure 1. A) A surface scan of the normal rat erythrocyte. Atomic-force microscopy, a scan size of 1 × 1 μm². B) The scan cross-section profile of normal rat erythrocyte. The section line passes through the center of erythrocyte.

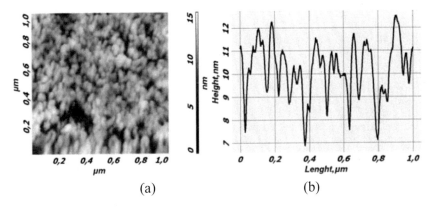

Figure 2. A) A surface scan of erythrocyte treated with noradrenaline. The specific concentration of the hormone was $2 \cdot 10^{-9}$ mol/mg protein. A scan of 1 × 1 μm². B) The scan cross-section profile of the rat erythrocyte surface treated with noradrenaline. The section line passes through the center of erythrocyte.

2.2. Investigation of Structural Transitions in Biomembranes by Fluorescence Methods

The mechanism underlying the effect of noradrenaline on the surface morphology of erythrocyte membranes was discovered using fluorescence methods and IR spectroscopy. Erythrocyte ghosts of male Wistar rats were obtained after their hemolysis in a hypotonic phosphate buffer (pH 7.4) containing 2.75 mM KH_2PO_4 and 8.5 mM Na_2HPO_4. The ghosts were sedimented by centrifugation at 5500 g, and supernatant was decanted. The procedure was repeated four times. The ghosts were obtained and stored at 4°C. Fluorescence measurements were carried out on a Shimadzu RF-5301(PC)SCE spectrofluorimeter. 3 mL of a hypotonic phosphate buffer containing 2.75 mM KH_2PO_4 and 8.5 mM Na_2HPO_4, pH 7.4, and erythrocyte ghosts were poured in a $1 \times 1 \times 4$ cm^3 quartz cuvette. The concentration of protein ghosts, which was determined by the Warburg–Christian method from changes in the optical density of suspension [31], varied within 0.100-0.250 mg/mL, on the average. The specific concentration of noradrenaline was calculated as the concentration of hormone divided by the concentration of protein ghosts in suspension.

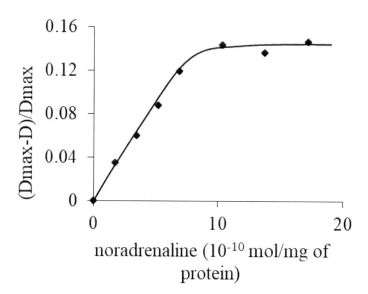

Figure 3. The dependence of $Q_1 = \dfrac{D_{max} - D}{D_{max}}$ on the specific concentration of noradrenaline added into the cuvette.

The cuvette with the ghost suspension was placed in the spectrofluorimeter thermostat for 10 min. The cuvette temperature reached a steady-state mode, which was controlled with an electronic thermometer. In all the experiments, the cuvette temperature was 36°C. After establishing a steady-state temperature mode in the cuvette, control measurements were made. Figure 3 displays the dependence of $Q_1 = \dfrac{D_{max} - D}{D_{max}}$ on the specific concentration of noradrenaline introduced into the cuvette, where D_{max}, D are the absorption intensities of tryptophan in ghosts at $\lambda = 228$ nm without hormone addition and with the hormone, respectively. The concentration of membrane protein was $C = 0.124$ mg/mL. The excitation wavelength was $\lambda = 332$ nm. Relative measurement error Q_1 was equal to 6%.

A noradrenaline solution with the concentration 10^{-6} M was prepared in a hypotonic phosphate buffer. The relative measurement error was 6%. The binding constant K_C was calculated for the hormone by the method reported in [32]; the stoichiometric concentration of the bound hormone B_{max} and a change in free energy of the system ΔG were also calculated.

The interaction of noradrenaline with erythrocyte membrane resulted in manifestation of the hypochromic effect (Figure 3). This testifies that structural orderliness of membrane proteins increases due to the tangle → β-structure and tangle → α-helix transitions. It is known that deterioration of absorption may be related to a change in the direction of dipole moments of quantum transitions of the monomeric residues of proteins that accompanies their transition to another conformation [33]. The effect of noradrenaline increases the biomembrane orderliness.

The absorption measurements were used to calculate K_C, B_{max} and ΔG for noradrenaline (Table 1). The same binding parameters were measured also for other hormones [22-24]. It is seen (Table 1) that male sex hormones (androsterone and testosterone) and stress hormones (cortisol, adrenaline and noradrenaline) have a high affinity to erythrocyte membrane.

Table 1. Parameters of hormone binding to erythrocyte membrane estimated from intrinsic tryptophan fluorescence quenching in membrane proteins

Steroid hormone	Binding constant K_C (M^{-1})	Amount of bound hormone B_{max} (mol/mg protein)	A change in Gibbs free energy ΔG (kJ/mol)
cortisol	$(1.23 \pm 0.12) \cdot 10^6$	$(4.69 \pm 0.47) \cdot 10^{-10}$	−36.0
noradrenaline	$(1.70 \pm 0.17) \cdot 10^6$	$(8.10 \pm 0.81) \cdot 10^{-10}$	−36.9
adrenaline	$(6.3 \pm 0.63) \cdot 10^6$	$(1.60 \pm 0.16) \cdot 10^{-11}$	−40.2
testosterone	$(2.24 \pm 0.22) \cdot 10^6$	$(1.09 \pm 0.11) \cdot 10^{-9}$	−37.6
androsterone	$(3.2 \pm 0.32) \cdot 10^6$	$(9 \pm 0.45) \cdot 10^{-11}$	−38.5
DHEA	$(5.99 \pm 0.60) \cdot 10^4$	$(1.80 \pm 0.18) \cdot 10^{-8}$	−28.3
DHEAS	$(1.56 \pm 0.16) \cdot 10^4$	$(4.03 \pm 0.40) \cdot 10^{-8}$	−24.8

The highest values of the binding constant and changes in free energy ΔG were observed for these hormones. Therewith, the amount of the bound hormone B_{max} necessary to produce maximum changes in the membrane conformation had the lowest value. The second group of hormones includes dehydroepiandrosterone (DHEA) and dehydroepiandrosterone sulfate (DHEAS). Their binding constants are approximately by two orders of magnitude lower as compared to androsterone. A change in free energy ΔG for these hormones is lower almost by 30%. The amount of the bound hormone B_{max} that stops conformational changes of proteins is higher by two-three orders of magnitude. Thus, the stronger is the hormone affinity to the membrane (Table 1), the greater is the hypochromic effect.

When the hormone concentration is higher than B_{max}, the fluorescence quenching virtually does not increase. The hormones tested in our study can bind to plasmatic membrane both specifically (to the membrane receptors) and nonspecifically (to molecules of the lipid bilayer). A static case of fluorescence quenching of tryptophan groups in membrane proteins is implemented [47]. When all specific binding sites are occupied by the hormone, a further raising of the hormone concentration in suspension produces only a slight increase in the quenching because nonspecific binding of the hormone to molecules of the phospholipid bilayer weakly changes the conformation of membrane proteins [47].

Microviscosity of erythrocyte membranes was measured on a Shimadzu RF-5301(PC)SCE spectrofluorimeter. The experimental specimen was prepared by the following procedure. 3 mL of a hypotonic phosphate buffer (pH 7.4) containing 2.75 mM KH_2PO_4 and 8.5 mM Na_2HPO_4, erythrocyte ghosts, a specified amount of nanoparticles or hormones, and a fluorescence pyrene probe were placed in a 1×1×4 cm^3 quartz cuvette. Prior to the experiment, all the components were stored at 4°C. The concentration of protein ghosts in the cuvette was 0.050 – 0.250 mg/mL, and that of pyrene, $7.76 \cdot 10^{-6}$ M. Pyrene was diluted in ethanol, its initial concentration was $1.5 \cdot 10^{-3}$ M. The cuvette was placed in the spectrofluorimeter thermostat for 10 min, and then the fluorescence measurements were made at 36°C. Before loading the specimen in the spectrofluorimeter thermostat, it was vigorously shaken for 1 min. To measure the fluorescence of ghosts loaded with a different amount of nanoparticles, a new specimen was prepared each time by the same procedure. Such a procedure is necessary because pyrene facilitates a rapid degradation of erythrocyte membranes.

Microviscosity of the lipid bilayer near membrane proteins (the region of protein-lipid interaction) was measured using the excitation wavelength $\lambda = 281$ nm and spectral slit width 1.5/5. Microviscosity of the lipid bilayer at a distance from membrane proteins (the region of lipid-lipid interaction) was measured at the excitation wavelength $\lambda = 337$ nm and spectral slit width 1.5/3. The emission maxima were observed at $\lambda = 374$ nm and $\lambda = 393$ nm (the vibronic peaks of excited pyrene monomers) and $\lambda = 468$ nm (the emission maximum of pyrene dimers). The concentration of ghosts was C = 0.128 mg protein/mL. The pyrene concentration in suspension was $7.7 \cdot 10^{-6}$ M; the temperature of suspension, 309.1 ± 0.1 K (36°C); and pH of suspension, 7.4 (Figure 4).

Relative microviscosity of the membranes was described by the relationship $L = \dfrac{\eta(A)}{\eta(0)}$, where $\eta(A)$ and $\eta(0)$ are the membrane microviscosities after the introduction of noradrenaline with the concentration A into suspension and without the hormone addition, respectively. For the region of lipid-lipid interaction, relative microviscosity L was calculated by the formula

$$L = \frac{\eta(A)}{\eta(0)} = \frac{F_{468}(0)}{F_{468}(A)} \cdot \frac{F_{393}(A)}{F_{393}(0)}$$

where $F_{468}(A)$ is the fluorescence intensity of pyrene at $\lambda = 468$ nm at the hormone concentration A in suspension; $F_{468}(0)$ is the fluorescence intensity of pyrene at $\lambda = 468$ nm in the absence of hormone in suspension. $F_{393}(A)$ and $F_{393}(0)$ are the fluorescence intensities of pyrene at $\lambda = 393$ nm and the hormone concentration A in suspension and in the absence of hormone, respectively. For the region of protein-lipid interaction, relative microviscosity L was calculated by the formula

$$L = \frac{\eta(A)}{\eta(0)} = \frac{F_{468}(0) - I_{468}}{F_{468}(A) - I_{468}} \cdot \frac{F_{393}(A) - I_{393}}{F_{393}(0) - I_{393}}$$

where I_{393} and I_{468} are the fluorescence intensities of tryptophan residues in membrane proteins at $\lambda = 393$ nm and $\lambda = 468$ nm, respectively.

Microviscosity measurements at each concentration of the hormone in suspension were performed at least three times. The average value was calculated in each point, and the measurement errors were estimated. The relative measurement error of relative microviscosity was equal to 6%.

Figure 4 displays the dependence of relative microviscosity L of membranes on the specific concentration of noradrenaline introduced into the cuvette. Dotted line corresponds to changes in relative microviscosity in the region of lipid-lipid interaction, and solid line – to changes in relative microviscosity in the region of protein-lipid interaction.

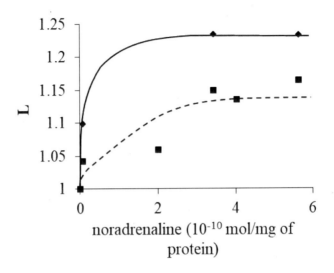

Figure 4. The dependence of relative microviscosity L of membranes on the specific concentration of noradrenaline added into the cuvette.

The greatest increment of relative microviscosity under the action of noradrenaline reaches 25% in the region of protein-lipid interaction. As the hormone concentration is raised, microviscosity increases, and then the saturation occurs at a certain concentration. The behavior of microviscosity correlates with changes in the absorption intensity of tryptophan in ghosts (Figure 3) that are caused by changes in the conformation of membrane proteins. Under the action of noradrenaline, the microviscosity curve reaches a constant value at $1 \cdot 10^{-9}$ mol/mg protein. The dependence of Q_1 for tryptophan in membrane proteins at the same concentrations of the hormone reached a maximum (Figure 3). Thus, the data obtained testify to the cooperative nature of changes in erythrocyte membranes subjected to the action of noradrenaline.

Microviscosity in the region of protein-lipid interaction, unlike microviscosity in the region of lipid–lipid interaction, starts to increase at lower concentrations of hormones, and the increment is more pronounced (Figure 4). This allows a conclusion that exactly the membrane proteins initiate the structural transitions in biomembranes that change the state of biomembranes.

2.3. The Mechanism of the Hormone Effect on Structural Transitions in Biomembranes

A large group of hormones gives rise to similar structural transitions in biomembranes. The mechanisms underlying the effect of hormones on structural transitions in biomembranes were studied earlier [22-24].

Cortisol is known to be a hydrophobic compound that is difficultly soluble in water. However, its structure includes three OH groups (in the 11, 17 and 21st positions) and two keto groups (in the 3 and 20th positions). According to IR spectroscopy data [22 - 24], exactly these groups are involved in the formation of hydrogen bonds with CO and NH groups of membrane-bound proteins and phospholipids. Cortisol increases intensity of the absorption band of CO bond approximately by 20%, which indicates an increase in the orderliness in membrane proteins due to the structural transitions tangle → α-helix. In addition, there occurred shifts of stretching vibrations of the peptide NH bond (3308→ 3280 cm^{-1}) and CH bond (2948 → 2862 cm^{-1}) caused by the formation of hydrogen bonds between cortisol and NH groups of proteins.

As a result, cortisol is involved in the formation of protein-lipid domains in erythrocyte membranes, and mechanical compressive stresses are generated in these domains. Cortisol molecules do not penetrate deep into the membrane and do not change the conformational state of the spectrin-ankyrin-actin network. Water dipoles are displaced to the boundaries of these domains. This is promoted by the enhancement of hydrophobic interactions owing to the additional two CH$_3$ groups (in the 10 and 13th positions) and hydrophobic rings of the hormone. An excessive content of water on the domain boundaries and the emergence of mechanical tensile stresses in these regions lead to the formation of mesostrips on the erythrocyte surface, which loosen the membrane. Overall, stiffness of the membrane increases, especially in the region of protein-lipid interaction. The degree of structural ordering of membranes increases, and free energy of the system ΔG decreases (Table 1). Such erythrocytes cannot satisfactorily perform their functions.

Noradrenaline also has the active groups. It can penetrate deeper into the membrane by its hydrophobic ring as compared to cortisol. Having a higher affinity to the membrane (Table 1), noradrenaline creates a greater number of hydrogen and hydrophobic bonds in the membrane and has a higher constant of binding to the membrane than cortisol. IR spectroscopy data testify to a increased orderliness in membrane proteins due to structural transitions tangle → α-helix and tangle → β-structure.

Adrenaline has even a higher affinity to erythrocyte membranes (Table 1). The adrenaline molecule includes three OH groups able to participate in the formation of hydrogen bonds, the NH group with a positive charge, and the adjacent CH_3 group that increases the effect of this charge. Adrenaline can bind both specifically to their adrenoreceptors and nonspecifically, being adsorbed on the membrane surface. IR spectroscopy [22 - 24] revealed the enhancement and formation of new hydrogen and hydrophobic bonds between hormones, membrane proteins and phospholipids, which led to the formation of protein-lipid domains. Normal compressive stresses increased in the membrane. They were supplemented with compression of the spectrin-ankyrin-actin network, which appeared upon interaction of adrenaline with its membrane adrenoreceptors.

Testosterone, being more hydrophobic than cortisol, penetrated deeper into the membrane. Its binding constant to erythrocyte membrane was also higher as compared to cortisol (Table 1). IR spectroscopy data [22 - 24] indicate that the intensity of absorption bands at 1544, 1656 and 3292 cm^{-1} increased by 30% and more. The 3308 → 3272 cm^{-1} ($\Delta v = 36$ cm^{-1}) shift of the absorption band of the NH bond was observed. The absorption bands at 2852 and 2932 cm^{-1} grew in intensity, and the intensity ratio of these bands changed. The increase in the integrated intensity of these absorption bands testifies to an increase in the orderliness of membrane proteins and, in particular, a growth of the fraction of α-helices. The increase in the fraction of α-helices occurs due to the structural transition tangle → α-helices. The 3308 → 3272 cm^{-1} ($\Delta v = 36$ cm^{-1}) shift of the absorption band of the NH bond is related to the formation of hydrogen bond between keto group ($C_3 =$ O) of the A ring and NH bond of the peptide group or the indole nucleus of tryptophan. The increase in intensity of 2932 and 2852 cm^{-1} bonds and the increase in the intensity ratio 2852/2932 cm^{-1} confirm the growing orderliness in the entire membrane. The absorption band at 1740 cm^{-1} (the C = O bond of the phospholipid ester group) grew in intensity and shifted to the short-wave region. The increased intensity of the C = O bond reflects the growing orderliness of phospholipids in the domains and domains between each other. The short-wave shift of this band is caused by the formation of hydrogen bond between OH group at the C_{17} carbon atom in the D ring of testosterone and C = O bond of phospholipids. The 1088 → 1098 and 1236 → 1248 cm^{-1} band shifts to the short-wave region are related to dehydration of phospholipids caused by the growth of their orderliness, since hydration shifts these bands to the long-wave region. An increase in intensity of the bands at 1098 and 1247 cm^{-1} (P-O-C and P = O bonds of phospholipids, respectively) relative to control samples corroborates the increase in the orderliness of phospholipids under the action of the hormone.

Androsterone. The incubation of rat erythrocyte ghosts with androsterone (C = 2.76·10^{-8} M) shifted the frequency of NH bonds [22 - 24]. A 30% and greater increase in the integrated intensity of absorption bands of CO (1654 cm^{-1}) and NH (3280 cm^{-1}) bonds was observed. The absorption band corresponding to the β-structure appeared at 1635 cm^{-1}. Stretching vibrations of CH bonds at 2848 and 2930 cm^{-1} grew in intensity. The frequency shift of the NH bond of membrane peptides is related to the formation of hydrogen bond with the C = O group of the hormone D-ring. All this increased the binding constant in comparison with cortisol (Table 1) and caused more pronounced structural transitions in the membrane. The increase in intensity of the CO-peptide bond was related to an increase in the content of α-helices owing to the tangle → α-helix transition in membrane proteins. The increase in intensity of the absorption band 1620 – 1635 cm^{-1} was produced by the structural transition of the tangle → β-structure type in membrane proteins.

DHEA. The incubation of DHEA with erythrocyte ghosts showed [22 - 24] that the frequency of stretching vibrations of the NH-peptide bond shifted by 20 cm^{-1} (3308 → 3288 cm^{-1}) to the long-wave region. The integrated intensity of the bands at 1546, 1654.9 and 3288 cm^{-1} increased. The band shift 1236 → 1247.6 cm^{-1} reflects dehydration of phosphate groups in phospholipids. The 1748 → 1732 cm^{-1} frequency shift of the C = O bond of phospholipids was observed along with an increase in the band intensity. The shifts of band frequency 2930 → 2925.8 and 2848 → 2851.8 cm^{-1} (CH stretching vibrations) occurred along with an increase in intensity. A change was observed in the intensity ratio 2852/2924 cm^{-1}. This testifies to the enhanced orderliness of membrane phospholipids. The binding constant of DHEA is much lower as compared to cortisol (Table 1).

DHEAS. The incubation of DHEAS with erythrocyte ghosts [22 - 24] produced a shift of the absorption band 3308 → 3286 cm^{-1} ($\Delta v = 22$ cm^{-1}) (NH-peptide bond). The bands at 1548, 1656 and 3298 cm^{-1} grew in intensity relative to the control sample, but this increase was less pronounced than in the case of DHEA. The absorption bands at 1632 and 1684 cm^{-1} assigned to the β-structure were observed. The band shifts 2930 → 2928 and 2848 → 2852 cm^{-1} ($\Delta v = 4$ cm^{-1}) were recorded along with a change in the 2852/2928 cm^{-1} ratio. The band at 1236 cm^{-1} (the P=O bond) showed a strong splitting and had 3-4 bands in the range of 1236-1256 cm^{-1}. The bands at 1084 and 1100 cm^{-1} (the P–O–C bond) were observed. The 1748 → 1738 cm^{-1} shift was smaller than that in the specimen with DHEA.

Overall, it can be concluded that the interaction of DHEAS with erythrocyte ghosts leads to the formation of hydrogen bonds between keto group ($C_{17} = O$) and NH peptide bonds, and also between OH group at C_3 in the A ring and C = O group of phospholipids. The formation of the indicated hydrogen bonds results in the ordering of membrane proteins and phospholipids (the tangle → α-helix transition). The hydrophobic interaction of the hormone with the surface of erythrocyte ghosts also makes some contribution to the structural changes of membranes. Upon interaction with DHEAS, this effect is observed but becomes less pronounced. The substitution of OH group by SO_3 strongly decreases the hydrogen bond energy. All this decreased the binding constant of DHEAS to the membrane (Table 1). The lowest binding constant is caused by the highest hydrophilicity of the hormone, which prevents the hormone from penetrating deep into the phospholipid bilayer.

The analysis of IR spectra of the ghosts after their incubation with hormones allows a conclusion on the formation of new hydrogen bonds between molecules in membrane, the occurrence of tangle → α-helix and tangle → β-structure transitions in membrane proteins, and an increase in the orderliness of phospholipids. This increases microviscosity of membranes and changes the field of mechanical stresses in the membrane. In its turn, this leads to structural rearrangements in the membrane, and changes the erythrocyte shape and its surface morphology.

2.4. The Features and Structure of Protein-Lipid Domains in Plasmatic Membranes

Which types of protein-lipid domains exist in plasmatic membranes? Rafts, the regions of the lipid bilayer enriched with cholesterol and sphingolipids. Their composition determines their structure and properties. These regions intussuscept some membrane proteins, for example, caveolins. Rafts with caveolin are involved in endocytosis and exocytosis [34, 35]. However, our experiments demonstrated that protein-lipid domains can emerge also near the membrane proteins bound to cytoskeleton.

First, changes occur in the conformation of membrane proteins and in the protein-lipid interactions (Figure 3).

Second, androgens and catecholamines produce the most pronounced changes in microviscosity of the lipid bilayer in the region of protein-lipid interaction. The greatest increase in the density of lipids is observed near proteins (Figure 4).

Third, the appearance of folds on plasmatic membranes subjected to the action of androgens and catecholamines indicates the enhancement of mechanical stresses in the spectrin-ankyrin-actin network (Figures 1 and 2). This issue will be discussed below.

And fourth, upon addition of cytochalasin D to the suspension of erythrocytes with noradrenaline, folds on the membrane surface were not observed [36, 37]. It means that stresses are not generated in the membrane without conformational changes of membrane and cytoskeleton proteins.

It can be concluded that stable protein-lipid domains are formed in plasmatic membranes under the action of hormones (androgens and catecholamines). They are formed around membrane proteins bound to cytoskeleton. In distinction to the rafts, which have a short lifetime, these domains exist for a longer time until hormones remain bound to plasmatic membrane.

2.5. The Effect of Structural Transitions in Biomembrane on the Activity of Na$^+$, K$^+$-ATPases in Erythrocyte Membranes

The effect of structural transitions in biomembrane on its function was exemplified by their effect on the activity of Na$^+$, K$^+$-ATPases in erythrocyte membranes. The activity of Na$^+$, K$^+$-ATPases in erythrocyte membrane was estimated by the method reported in [38], which is based on accumulation of inorganic phosphorus (P$_i$) in a medium containing ATP. The essence of the method is that in an acid medium ammonium molybdate and phosphoric acid (and its salts) after the hydrolysis by ATP enzyme enter the reaction to yield ammonium phosphomolybdate. The reduction of the latter gives a mixture of different molybdenum oxides having a blue color.

20 μL of the suspension of erythrocyte membranes with the concentration C = 20 mg protein/mL was supplemented with hormones and 40 μL of the incubation medium of the following composition (mM): NaCl – 125, KCl – 25, MgCl$_2$ – 3, EDTA – 5, ATP – 2, and Tris-HCl – 50 (pH 7.35). The concentration of ghosts roughly corresponded to their concentration in blood. As a control, the same composition in another vial was supplemented with 5 μL of ouabain (MP Biomedicals, LLC) dissolved in a hypotonic Tris-HCl buffer with the concentration of 10^{-2} M. The incubation was performed at 37°C for 1 h. The reaction was stopped by 40 μL of a 20% trichloroacetic acid solution. The precipitated protein was separated by centrifugation at 330 g for 15 min. To determine the amount of the produced inorganic phosphorus (P$_i$), 40 μL of supernatant in each plate well was supplemented with 100 μL of the molybdenum reagent (2.5 g (NH$_4$)$_6$Mo$_7$O$_2$, 13 mL H$_2$SO$_4$ conc., 200 mL H$_2$O dist.) and 40 μL of a 7% aqueous solution of ascorbic acid. After 20 minutes, the absorption difference at 630 and 495 nm was measured photometrically in a STAT

FAX-2100 (Awareness Technology Inc.) microplate photometer. The concentration of inorganic phosphorus in each specimen and the enzymatic activity were calculated from calibration curves. The activity of Na^+,K^+-ATPase was obtained as the difference between activity of the enzyme measured without ouabain and with ouabain. The activity was expressed in µmol/h·mg protein.

The measurement errors appeared due to the errors in volumetric batching of the ghost suspension specimens and their titration with hormones. The relative measurement error was 3% for the absorption values, and 6% for the activity of Na^+,K^+-ATPase.

The dependence of Na^+,K^+-ATPase activity in erythrocyte membranes on the concentration of noradrenaline in suspension was measured (Figure 5). The dependence was plotted versus the specific concentration of hormones: the concentration of hormones divided by the concentration of ghost proteins in suspension. It was shown that the enzyme activity first increased with raising the concentration of hormones. After reaching a certain maximum value, it started to decrease. Under the action of noradrenaline, the activity maximum of Na^+, K^+-ATPases in erythrocyte membranes was observed at the specific concentration of $0.5 \cdot 10^{-10}$ mol/mg protein (Figure 5). The maximum activity under the action of noradrenaline was four time higher than in the absence of the hormone.

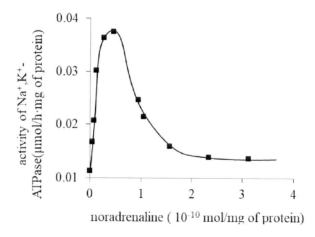

Figure 5. The activity of Na^+, K^+-ATPases in erythrocyte membranes versus the concentration of noradrenaline in suspension.

The dependence of activity of Na^+, K^+-ATPases in erythrocyte membranes on specific concentration of hormones correlated with changes of their microviscosity under the action of hormones in suspension (Figure 4). Microviscosity of membranes and activity of Na^+, K^+-ATPases increased in parallel, over the same range of specific concentrations of hormones in suspension. The peak in the enzyme activity corresponded to the hormone concentration that was lower than the concentration at which microviscosity curve reached a constant value (Figure 4). After that, deterioration of the enzyme activity was noted. Similar measurements were made also for other hormones. The greater was the increase in microviscosity of erythrocyte membrane under the action of the hormone, the stronger was the increase in the activity of Na^+,K^+-ATPases. For DHEA, a change in microviscosity was below the critical value and did not affect the activity [26, 28].

2.6. The Effect of Mechanical Stresses on the Activity of Na^+, K^+-ATPases in Erythrocyte Membranes

The results obtained make it possible to understand changes in the activity of Na^+, K^+-ATPases in erythrocyte membranes. The behavior of the enzyme activity correlated with the dependence of microviscosity on the concentration of hormones in suspension. Androsterone and testosterone increased membrane microviscosity to a greater extent [22-24]. The same hormones produced a greater increase in the enzyme activity [26, 28].

Stability of the complex of proteins and lipids in biomembrane is caused by both the hydrophobic and electrostatic interactions. According to this model, mutual displacements of enzyme subunits in the protein-lipid complex of Na^+, K^+-ATPases and lipids would result in deformation of the lipid bilayer. The protein loop with the binding sites of ions would move through the membrane, transferring three sodium ions out from the cell, and two potassium ions into the cell [39, 40, 41]. A part of energy generated by ATP hydrolysis is consumed in the process. Hence, a considerable increase in the membrane stiffness will increase this part of energy and deteriorate the enzyme activity. Na^+, K^+-ATPase takes energy for its operation from the hydrolysis of ATP, but the jump of the binding site requires mechanical vibrations of the enzyme subunits and the entire cell with a certain frequency. This was established in the experiments [42] where spectral frequencies of mechanical vibrations of non-native and native cells were compared with the operating Na^+, K^+-ATPases.

Among the driving forces of conformational transformations of enzyme are mechanical vibrations (phonons) that are transmitted to subunits from the lipid bilayer. Being not a single driving force of conformational transformations of the enzyme, they affect the enzyme activity. As microviscosity (stiffness) of the membrane is increased, the phonon energy of the bilayer also increases. This can be a cause of the primary growth of enzyme activity. As stiffness is further increased, the enzyme activity decreases (the dome-like dependence on the concentration of hormones in suspension).

Let us consider the membrane as a liquid crystal. Notions of the theory of solids can be applied to the properties of biomembranes [43, 44]. According to the Debye theory [45], the maximum energy of phonons in a solid represented by a cube of volume L^3 is equal to

$$\hbar\omega_{max} = 2\pi \cdot \langle v \rangle \cdot \left(\frac{3N_A}{4\pi L^3}\right)^{1/3}$$

where N_A is the Avogadro number, $\hbar$ – the reduced Planck's constant, ω_{max} – the maximum frequency of normal vibrations of the set of body atoms, and $\langle v \rangle$ – the average sound velocity determined by the relationship

$$\frac{3}{(\langle v \rangle)^3} = \frac{1}{v_\ell^3} + \frac{2}{v_t^3}$$

where v_ℓ and v_t are the velocities of longitudinal and lateral modes, respectively.

As the concentration of hormones in erythrocyte suspension is raised, the density of their membranes increases, thus increasing the membrane microviscosity. It was found in experiments that an increase in density of a solid increases the average sound velocity in this solid and the maximum energy of phonons [45]. This is why loading of membranes with androgens and stress hormones increases the maximum energy of their phonons. The higher is the phonon energy of the membrane, the easier is the jump of cations. Thus, an increase in microviscosity exerts a double effect on the enzyme activity. At first, an increase in microviscosity raises the energy of phonons transmitted from the lipid bilayer to the enzyme subunits. This enhances its activity. A further growth of microviscosity hinders the conformational transitions in enzyme, the ATP delivery to the enzyme, and the ATP hydrolysis. Activity of the enzyme starts to decrease. When the concentration of hormones is raised from the zero value, at first an increase in activity is predominant, and then – a decrease.

This hypothesis is supported by the fact that without lipid bilayer the activity of Na^+, K^+-ATPases strongly decreases, and the removal of the spectrin-ankyrin-actin network from the membrane decreases the enzyme activity more than twofold [38]. In the absence of lipid bilayer, elastic vibrations cannot be transferred to Na^+,K^+-ATPases.

The activity of Na^+, K^+-ATPases affects the surface electric potential of erythrocyte membranes. The higher the activity, the greater the zeta-potential. This observation will be useful for explaining the nature of erythrocyte stasis in microcapillaries. In [27], we studied the perfusion of the isolated rat heart. Erythrocytes in the perfusate were pretreated with adrenaline and then washed in a phosphate buffer to remove unbound hormones from the suspension. This allowed us to decrease the heart rate in 5 min and sharply reduce the amplitude of cardiac contractions. In 7 min the heart stopped. Dark injury zones were observed on the pink surface of the left ventricle.

The cardiac arrest was caused by acute myocardial hypoxia. The results of these experiments can be interpreted as follows. Erythrocytes treated with a large amount of adrenaline increased stiffness of their membranes and could not move in the heart microcapillaries. In addition, the increased microviscosity of membranes deteriorated the activity of Na$^+$,K$^+$-ATPases, which was followed by a decrease in the zeta-potential of membranes. Erythrocytes started to agglutinate into rouleaux (slugging). This led to hemostasis in the cardiac muscle and then to myocardial hypoxia. This occurred exactly due to structural changes of erythrocytes membranes because the addition of adrenaline into solution (without erythrocytes) around the heart prevented cardiac hemostasis.

Similarly, hemostasis, acute myocardial hypoxia, and even death can occur in young sportsmen during training or competition if they use steroid hormones [46].

2.7. Mechanical Stresses in Plasmatic Membranes

The previous sections presented experimental data that prove the effect of mechanical stresses in the lipid bilayer on the functions of membranes. Let us consider the theoretical models describing such stresses. A mature normal erythrocyte is shaped as a biconcave discocyte with alternating convex and concave regions of the membrane. Figure 6 displays a cross-section of a healthy erythrocyte with a discocyte shape. The radius of convex regions of the membrane is denoted as R_1. The radius of concave regions is R_2. The internal space of erythrocyte is denoted as in, and the space outside the cell, as out. Which mechanical stresses exist in these regions?

Direct experimental data on the subject are lacking. The following model can be proposed (Figure 7). The spectrin-ankyrin-actin network consists of the protein filaments in which compressive stresses occur. The attaching points form a network of trigonal cells with the 100 nm side [47]. The network creates the pressure on the inner side of membrane, which is directed inside the cell. The value of such stresses can change. This network, being attached to the membrane, creates a radial external pressure P_c applied to the inner surface of membrane (from hemoglobin) and directed to the cell center. The pressure normal to the membrane surface can be roughly presented as the sum of the constant and the quantity with periodical coordinates.

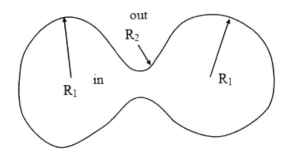

Figure 6. Cross-section of a healthy erythrocyte having the discocyte shape.

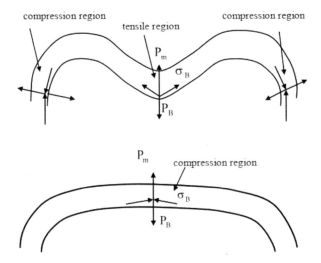

Figure 7. Changes in the curvature of erythrocyte membrane upon adsorption of noradrenaline molecules on the membrane.

In addition, there is a pressure difference between internal and external media of the cell, $\Delta P = P_{in} - P_{out}$, which is caused by the difference between hydrodynamic and osmotic pressures of liquid in the cell, P_{in}, and outside it, P_{out}. Under usual conditions, P_{in} slightly exceeds P_{out}, so that $\Delta P > 0$ and the cell swells. Swelling of the cell is prevented by the pressure of the spectrin-ankyrin-actin network P_c that has the opposite direction, inside the cell. Supposedly, under normal conditions P_c is slightly higher than ΔP, and their difference $P_B = P_c - \Delta P$ should be compensated by the radial resultant force P_m of internal stresses σ_B acting on the membrane element in the longitudinal direction (Figure 7).

Figure 7 illustrates changes in the erythrocyte membrane curvature upon adsorption of noradrenaline molecules on the membrane. Biconcave discocyte (top image) becomes biconvex due to the increased compressive stress in the membrane (bottom image).

These suppositions are supported by the following facts. As the osmotic pressure of the external medium decreased and the medium became hypotonic, $\Delta P > 0$ became very high, the cell swelled, and then the membrane burst (hemolysis of erythrocytes). Hemoglobin leaked to the external medium. Cytoskeleton was not damaged, but erythrocyte lost its shape, looked like a blown off air ball and turned into a ghost. This testifies that the cytoskeleton itself cannot retain normal shape of the cell. As the osmotic pressure of external medium increased, the medium became hypertonic, $\Delta P < 0$, and the cell was compressed, turning into echinocyte.

As will be shown below, this model implies that compressive stresses prevail in the membrane. Only such stresses can ensure a long life of the membrane since the prevalence of tensile stresses results in destruction of the membrane.

The interaction of stress hormones and androgens with cell membranes is accompanied by structural ordering of membrane proteins (Sections 2.2, 2.3 and 2.4). The volume always decreases when going from any disorder to order; thus, the transition from disordered tangle to α-helix or β-structure decreased the volume of proteins. Hydrophobic interactions between proteins and adjacent lipids are enhanced with increasing the fraction of α-helices in membrane proteins [48]. In addition, the interaction of hormones with biomembrane increased the intensity of absorption band at 1740 cm^{-1} (C = O bond of the ester group in phospholipids) and shifted it to the short-wave region. The increased intensity of C = O bond reflects an increase in the orderliness of membrane phospholipids. The short-wave shift of this band is caused by the formation of the hydrogen bond between OH group of hormones and C = O bond of phospholipids. The short-wave shifts 1088 → 1098 and 1236 → 1248 cm^{-1} are related to dehydration of phospholipids caused by the increase in their orderliness, because hydration shifts these bands to the long-wave region. An increase in intensity of the bands at 1098 and 1247 cm^{-1} (P-O-C and P = O bonds of phospholipids, respectively) with respect to control samples confirms that the orderliness of phospholipids increases under the action of hormones (Sections 2.2 – 2.4). Proteins and adjacent lipids formed the domains in which density of lipid molecules and hydrostatic pressure were increased and molecules of membrane water were displaced to the tensile zones at the domain periphery. On the domain boundaries, water molecules increased the distance between lipid molecules, which created tensile forces on the boundaries.

The indicated four factors – the decrease in the volume of membrane proteins, the enhancement of hydrophobic interaction between proteins and adjacent lipids, the ordering of lipids upon their interaction with hormones, and the displacement of water molecules to the domain periphery – created bulk forces in the membrane that strained and compressed the membrane. According to experimental data, it can be concluded that these bulk forces do not concentrate only near membrane proteins but act in the entire membrane. The attractive forces are created not only by the membrane proteins but also by the lipids interacting with hormones, and tensile forces are created by water molecules displaced to the periphery of the protein-lipid domains. The current expressions for components of the bulk force should significantly differ from zero at large distances from ankyrins.

Cytoskeleton prevented free expansion or compression of the membrane, which increased internal mechanical stresses in the membrane. An increase in microviscosity was most pronounced in the region of protein-lipid interaction, which corresponded to a decrease in the distances between membrane molecules residing close to the membrane protein. This increased the membrane density and mechanical compressive stresses in the membrane near membrane proteins bound to the spectrin-ankyrin-actin network (Sections 2.2 – 2.4). The strengthening of hydrophobic interactions and Van der Waals forces between membrane molecules also increased the longitudinal compressive stresses in the entire membrane, $\sigma_0 < 0$. If hormones did not interact with the spectrin-ankyrin-actin network and did not increase compressive stresses of the network, the constant component of pressure on the internal boundary of the membrane did not change.

What form is taken by the bulk forces emerging in the membrane upon its interaction with hormones? According to experimental data, the incorporation of hormones into the membrane creates regions of additional mechanical stress around ankyrins – the proteins by which erythrocyte membrane is attached to cytoskeleton. The attaching points form a network of trigonal cells with the 100 nm side [46]. It can be concluded that the field of bulk forces is a periodic function of coordinates, which should transform into itself when Cartesian coordinates are turned by 120°, 240°, 360°, etc. Such a function can be presented as a Fourier series. To simplify the calculations, one can consider not the entire series, but roughly present the components of the bulk force acting on the elementary volume of membrane as the sum of several harmonic functions:

$$\begin{cases} f_x = \sqrt{3} \cdot f_0 \cdot \left(-\cos k\left(\frac{\sqrt{3}}{2}x + \frac{y}{2} \right) + \cos k\left(\frac{\sqrt{3}}{2}x - \frac{y}{2} \right) \right) \\ f_y = f_0 \cdot \left(2\cos ky - \cos k\left(\frac{\sqrt{3}}{2}x + \frac{y}{2} \right) - \cos k\left(\frac{\sqrt{3}}{2}x - \frac{y}{2} \right) \right) \end{cases}$$

where $f_0 = \dfrac{E \cdot A \cdot k^2}{1 - v^2}$, E is the Young modulus, v – the Poisson ratio, A – the amplitude of displacement vector, and k is the wavenumber. Figure 8 shows how bulk forces break the membrane into equiangular triangles with the side of $\dfrac{4\pi}{k\sqrt{3}}$. Either compression (+) or tension (–) occurs in each of the triangles. The compression region is formed around the membrane protein attached to the spectrin-ankyrin-actin network. Such proteins are marked by black circles.

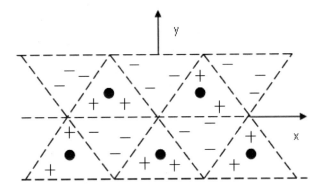

Figure 8. Compression and tensile regions in biomembrane.

Let us calculate tensors of mechanical stresses and displacements in the membrane during its interaction with hormones. The membrane is assumed to be a flat plate made of an isotropic material with thickness b. The OZ axis of Cartesian coordinates is perpendicular to the median plane of the membrane, OX and OY axes are directed along the plane. We think that components of the stress tensor

$$\sigma_{zz} = \sigma_{zx} = \sigma_{zy} \approx 0$$

are much smaller than longitudinal internal stresses in the membrane. The equilibrium equations:

$$\begin{cases} \dfrac{\partial \sigma_{xx}}{\partial x} + \dfrac{\partial \sigma_{xy}}{\partial y} = -f_x \\ \dfrac{\partial \sigma_{yy}}{\partial y} + \dfrac{\partial \sigma_{xy}}{\partial x} = -f_y \end{cases}$$

Components of the stress tensor:

$$\sigma_{xx} = \sigma_0 + \frac{EAk}{1-v^2} \cdot \left(-2v \cdot \sin ky + \left(\frac{3}{2} + \frac{v}{2} \right) \left(\sin k\left(\frac{\sqrt{3}}{2}x + \frac{y}{2} \right) - \sin k\left(\frac{\sqrt{3}}{2}x - \frac{y}{2} \right) \right) \right)$$

$$\sigma_{yy} = \sigma_0 + \frac{EAk}{1-v^2} \cdot \left(-2\sin ky + \left(\frac{3}{2}v + \frac{1}{2} \right) \left(\sin k\left(\frac{\sqrt{3}}{2}x + \frac{y}{2} \right) - \sin k\left(\frac{\sqrt{3}}{2}x - \frac{y}{2} \right) \right) \right)$$

$$\sigma_{xy} = \frac{\sqrt{3}}{2} \frac{EAk}{1+v} \cdot \left(\sin k\left(\frac{\sqrt{3}}{2}x + \frac{y}{2} \right) + \sin k\left(\frac{\sqrt{3}}{2}x - \frac{y}{2} \right) \right)$$

where $\sigma_0 = \text{const} < 0$ is caused by compression of the membrane due to increase in hydrophobic interactions and Van der Waals forces between membrane molecules.

Components of the deformation tensor:

$$u_{xx} = \frac{1-v}{E}\sigma_0 + \frac{3Ak}{2} \cdot \left(\sin k\left(\frac{\sqrt{3}}{2}x + \frac{y}{2} \right) - \sin k\left(\frac{\sqrt{3}}{2}x - \frac{y}{2} \right) \right)$$

$$u_{yy} = \frac{1-v}{E}\sigma_0 + Ak \cdot \left(-2\sin ky + \frac{1}{2}\left(\sin k\left(\frac{\sqrt{3}}{2}x + \frac{y}{2} \right) - \sin k\left(\frac{\sqrt{3}}{2}x - \frac{y}{2} \right) \right) \right)$$

$$u_{xy} = \frac{\sqrt{3}}{2} Ak \cdot \left(\sin k\left(\frac{\sqrt{3}}{2}x + \frac{y}{2} \right) + \sin k\left(\frac{\sqrt{3}}{2}x - \frac{y}{2} \right) \right)$$

$$u_{zz} = -\frac{2v}{E}\sigma_0 - \frac{2vAk}{1-v} \cdot \left(-\sin ky + \sin k\left(\frac{\sqrt{3}}{2}x + \frac{y}{2} \right) - \sin k\left(\frac{\sqrt{3}}{2}x - \frac{y}{2} \right) \right)$$

$$u_{xz} = u_{yz} = 0$$

Components of the displacement vector:

$$u_x = \frac{1-\nu}{E}\sigma_0 \cdot x + \sqrt{3}A \cdot \left(-\cos k\left(\frac{\sqrt{3}}{2}x + \frac{y}{2}\right) + \cos k\left(\frac{\sqrt{3}}{2}x - \frac{y}{2}\right)\right)$$

$$u_y = \frac{1-\nu}{E}\sigma_0 \cdot y + A \cdot \left(2\cos ky - \cos k\left(\frac{\sqrt{3}}{2}x + \frac{y}{2}\right) - \cos k\left(\frac{\sqrt{3}}{2}x - \frac{y}{2}\right)\right)$$

$$u_z = -\frac{2\nu}{E}\sigma_0 \cdot z - \frac{2\nu A k}{1-\nu} \cdot z\left(-\sin ky + \sin k\left(\frac{\sqrt{3}}{2}x + \frac{y}{2}\right) - \sin k\left(\frac{\sqrt{3}}{2}x - \frac{y}{2}\right)\right)$$

The results obtained describe the experimental data reported above (Sections 2.1–2.3). The analysis of components of the displacement vector revealed that homogeneous deformation and harmonic compression-tension of the membrane take place. Compression and tension zones have a staggered arrangement. In compression zones, phospholipids were displaced to the membrane protein bound to cytoskeleton. Figure 9 illustrates changes in the membrane structure upon loading with stress hormones and androgens: 1 – cytoskeleton filaments to which membrane proteins are attached; 2 –transmembrane proteins; 3 – membrane lipids; and 4 – water molecules. Top image – the unloaded membrane. Bottom image – the loaded membrane; compressive stresses appeared in the region of protein-lipid interaction, and density of lipids increased. Water molecules in the membrane are marked by circles.

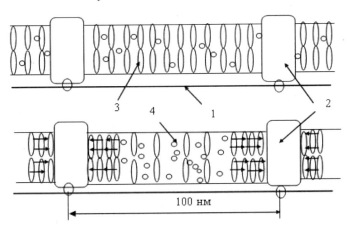

Figure 9. Structural changes of membranes upon loading with stress hormones and androgens.

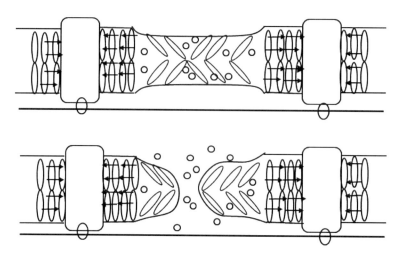

Figure 10. Top image – at a further increase in compressive stresses in the protein-lipid region, tensile stresses in the region of lipid-lipid interaction create a "neck." Bottom image – at a further increase in compressive stresses the membrane can burst.

In the unloaded state, they are uniformly distributed among lipid molecules. In the loaded state, they are displaced to the tensile region, thus increasing the distance between lipid molecules. The distance between transmembrane proteins to which cytoskeleton filaments are attached is ~100 nm [46]. Shifts of phospholipids are described by the components of the displacement vector. In these zones, the membrane thickness increased, which is described by the displacement vector (Figure 10). In local tensile zones, the membrane thickness decreased, which led to necking. Figure 10 (top image) demonstrates that at a further increase in compressive stresses in the protein-lipid region the tensile stresses in the region of lipid-lipid interaction create a "neck." The bottom image shows that at a further increase in compressive stresses the membrane can burst.

Some hormones, for example cortisol, which are adsorbed on the membrane surface but weakly interact with the spectrin-ankyrin-actin network, could promote the formation of cracks in the membrane [22, 24]. The necking and rupture of the membrane could occur if hormones did not affect the tension in spectrin-ankyrin-actin network, and distances between transmembrane proteins to which the network was attached remained constant. The transitions smectic A → smectic C (gel phase L_β → gel phase

$L_{\beta'}$) were likely to occur in the "necks" (tensile mesostrips) of the lipid bilayer (Figure 10). The membrane thickness decreased by 2 nm [22, 24]. At a further increase in tensile stresses, when they exceeded some critical value, the membrane could burst (Figure 10). Before the burst, the transition gel phase $L_{\beta'} \rightarrow$ liquid crystal phase L_α could occur in the tensile zones of lipid bilayer. In the process, density of the lipid bilayer decreased, which resulted in the formation of pores and then cracks.

Now it is possible to explain changes in the erythrocyte shape that occur upon loading with hormones or nanoparticles. On the internal boundary of convex regions of the membrane, normal compressive stresses $\sigma_B < 0$ are directed along the membrane boundary; only such stresses can generate the resultant pressure P_m that compensates the external radial pressure P_B (Figure 7).

A relation between stress in the membrane and compressive stress of cytoskeleton can be roughly described by the equilibrium equation for a spherical envelope with radius R_1 [49]:

$$\frac{2b \cdot \sigma_0}{R_1} = P_B$$

where b is the membrane thickness. As follows from this equation, an increase in $\sigma_0 < 0$ will increase the radius R_1 of convex regions of the membrane and the height of erythrocytes. This phenomenon was discovered when membranes were affected by hormones [22, 24]. The biconcave discocyte transformed into stomatocyte or biconvex discocyte.

When erythrocytes were loaded with nanocrystals whose diameters exceeded the critical value, a similar pattern was observed [29, 30]. The adsorption of nanocrystals with supercritical diameters increases the membrane microviscosity. This increases the membrane density and compressive stress in the entire membrane, $\sigma_0 < 0$.

2.8. The Effect of Mechanical Stresses in Membrane on the Erythrocyte Shape and Surface Morphology

Let us discuss the effect of mechanical stresses in membrane on the erythrocyte shape and its surface morphology. It will be demonstrated that the atomic force microscopy data can be interpreted as an increase in compressive stresses in the membrane upon its loading with stress hormones, androgens or nanoparticles.

The emergence of folds on the membrane surface (Figure 2) is affected mostly by the spectrin-ankyrin-actin network. This can be concluded from the fact that folding on the membrane surface appears under the action of the most hydrophobic hormones or the hormones that have receptors on the external surface of plasmatic membranes (adrenaline, androsterone, testosterone) [22, 24]. Direct experiments were carried out to reveal the effect of the spectrin-ankyrin-actin network on the formation of membrane folds. To verify the effect of noradrenaline on the structure of cytoskeleton, the study was performed with the cytoskeleton inhibitor – cytochalasin D, which stops the contraction of the spectrin-ankyrin-actin network [50]. Upon addition of cytochalasin D to the suspension of erythrocytes with noradrenaline, folds on the membrane surface were not observed [37].

The membrane behavior resembles the behavior of a loaded plate with hinged joints. The load is created by contraction of protein filaments of the spectrin-ankyrin-actin network upon interaction with hormones. Contractions of the network filaments generate stresses $P_B = P_c - \Delta P$ on the internal surface of membranes at the attaching points of the network (Figures 9 and 10). If the OX axis is drawn through the attaching points of contractive network to the membrane, the following form of pressure on the membrane from within the cell will be obtained (Figure 11). Figure 11 displays the pressure on the internal boundary of membrane versus the OX axis. Line 1 – P_c is the cytoskeleton pressure on the internal surface of the membrane; line 2 – $\Delta P = P_{in} - P_{out}$ is the pressure difference of liquid within the cell and outside it; and line 3 – $P_B = P_c - \Delta P$ is the pressure on the internal boundary of membrane.

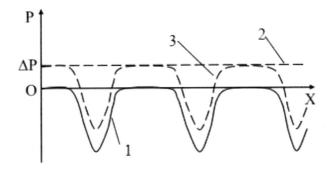

Figure 11. Pressure on the internal boundary of membrane versus the OX axis.

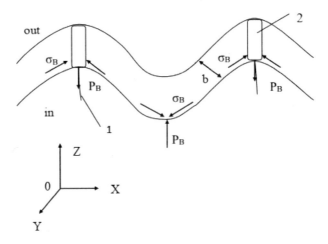

Figure 12. The distribution model of mechanical stresses on the boundaries of the convex region of membrane.

The membrane is assumed to be a plate of thickness b. The OZ axis of Cartesian coordinates is perpendicular to the median plane of the membrane, OX and OY axes are directed along this plane (Figure 12). Figure 12 displays a distribution model of mechanical stresses on the boundaries of the convex region of membrane. The membrane has thickness b. The internal volume of erythrocyte is denoted as 'in', and the external space as 'out'; 1 – cytoskeleton filaments that generate upon compression the P_c stress on the membrane surface directed to the cell center; 2 – transmembrane proteins to which the network filaments are attached. $P_B = P_c - \Delta P$ is the pressure on the internal boundary of membrane. $\Delta P = P_{in} - P_{out}$ is the pressure difference of liquid within the cell and outside it.

Let us analyze the folding parameters of erythrocyte membranes using the deformation model of a metal film on a viscous substrate subjected to thermoelastic stresses [51, 52]. The pressure on the internal boundary of membrane can be roughly presented as

$$P_B = P_0 \cdot \sin(kx)$$

where P_0 is a constant, $k = \dfrac{2\pi}{\lambda}$ is the wavenumber, and λ is the wavelength of the membrane folds.

The membrane folding emerges under the action of external pressure P_B and compressive stress in the membrane. In the model under consideration, the height of the membrane fold is not small in comparison with the membrane thickness b, but is much smaller that the size of erythrocyte:

$$\xi(x) = A \cdot \sin(kx)$$

where A is the amplitude of the membrane fold. The height of the membrane fold is equal to two amplitudes A, and the horizontal dimension of the fold is equal to the wavelength.

We think that components of the stress tensor

$$\sigma_{xy} = \sigma_{zx} = \sigma_{zy} \approx 0$$

are much smaller than the longitudinal normal internal stresses. The longitudinal stresses:

$$\sigma_{xx} = \sigma_0 = \text{const} < 0$$

As follows from the symmetry of boundary conditions,

$$u_{yy} = u_{xy} = u_{xz} = u_{yz} = u_{zz} = 0$$

The stress tensor in terms of the deformation tensor:

$$\sigma_{xx} = \frac{E(1-v)}{(1+v)(1-2v)} \cdot u_{xx} = \sigma_0$$

$$\sigma_{yy} = \frac{v}{1-v} \cdot \sigma_0$$

At a strong deflection of the membrane,

$$u_{xx} = \frac{\partial u_x}{\partial x} + \frac{1}{2}\left(\frac{\partial \xi}{\partial x}\right)^2$$

The substitution of deformation tensor into stress tensor gives the displacement vector:

$$u_x = C \cdot x - \frac{A^2 k}{8} \cdot \sin(2kx)$$

$$u_y = 0$$

$$u_z = \xi(x) = A \cdot \sin(kx)$$

where $C = \frac{(1+v)(1-2v)}{E(1-v)} \sigma_0 - \frac{1}{4} A^2 k^2$, E is the Young modulus, and v is the Poisson ratio. The analysis of expressions for the displacement vector components shows that the formation of folds in the membrane is accompanied by homogeneous compression and harmonic compression-tension. Regions of increased and decreased density of molecules can appear. This may result in the rupture of membranes.

The membrane bending ζ and stresses obey the Von-Karman equilibrium equations for a plate with the applied external forces [48]:

$$\begin{cases} D \cdot \Delta^2 \xi - b \cdot \dfrac{\partial}{\partial x_\beta}\left(\sigma_{\alpha\beta} \dfrac{\partial \xi}{\partial x_\alpha}\right) = P_B \\ \dfrac{\partial \sigma_{\alpha\beta}}{\partial x_\beta} = 0 \end{cases}$$

where $D = \dfrac{E \cdot b^3}{12(1-v^2)}$ is the cylindrical stiffness of the membrane; indices are summed with respect to α, β. The second equation is valid for all stress tensors under consideration. The first equation gives

$$\dfrac{A \cdot (2\pi)^2}{\lambda^2}\left(D\dfrac{(2\pi)^2}{\lambda^2} + b\sigma_0\right) = P_0 \qquad (1)$$

In the experiments reported in [22, 24], atomic force microscopy was used to measure the bending amplitudes of membrane A, which are equal to a half-height of the fold, and the wavelengths of the membrane folds λ, which are equal to the horizontal dimension of the fold and appear on the surface of rat membranes upon incubation with some hormones. The data obtained are listed in Table 2. They can be interpreted using formula (1).

The addition of DMSO to erythrocytes decreased compressive stresses σ_0 in the membrane and tensile stresses P_0 applied to the internal boundary of membrane by the spectrin-ankyrin-actin network. The fold wavelength λ increased, while the fold amplitude A decreased. Below, all parameters in Table 2 are compared to the case of erythrocytes treated with DMSO.

Table 2. Changes in the wavelength and amplitude of folds on the membrane surface upon loading with hormones

Rat erythrocytes	Length λ, nm	Amplitude A, nm	Specific concentration of hormone, mol/mg protein	
control	100	2-3		$\dfrac{\sigma_0 \cdot A}{\lambda^2} \sim P_0$
DMSO	200	0.5-1		$\dfrac{\sigma_0 \downarrow \cdot A \downarrow}{\lambda^2 \uparrow} \sim P_0 \downarrow$
cortisol	600	0.25	10^{-10}	$\dfrac{\sigma_0 \uparrow \cdot A \downarrow}{\lambda^2 \uparrow} \sim P_0$
noradrenaline	100	1-1.5	$2 \cdot 10^{-9}$	$\dfrac{\sigma_0 \uparrow \cdot A \uparrow}{\lambda^2 \downarrow} \sim P_0 \uparrow$
adrenaline	250 and 50	15 and 2.5	$5 \cdot 10^{-8}$	$\dfrac{\sigma_0 \uparrow \cdot A \uparrow}{\lambda^2 \uparrow} \sim P_0 \uparrow$
testosterone	400 and 50	10-12 and 5	$5 \cdot 10^{-9}$	$\dfrac{\sigma_0 \uparrow \cdot A \uparrow}{\lambda^2 \uparrow} \sim P_0 \uparrow$
androsterone	150	3-4	$5 \cdot 10^{-8}$	$\dfrac{\sigma_0 \uparrow \cdot A \uparrow}{\lambda^2 \downarrow} \sim P_0 \uparrow$
DHEA	220	10-12	$5 \cdot 10^{-9}$	$\dfrac{\sigma_0 \uparrow \cdot A \uparrow}{\lambda^2 \uparrow} \sim P_0$
DHEAS	100	1.5-2	$5 \cdot 10^{-9}$	$\dfrac{\sigma_0 \cdot A}{\lambda^2} \sim P_0$

Cortisol did not penetrate deep into the membrane; when increasing σ_0, it did not increase P_0. The fold amplitude A decreased, whereas the fold wavelength λ increased. The membrane surface became smoother. Large protein-lipid domains were formed, and cracks could emerge on their boundaries.

Noradrenaline, similar to cortisol, did not penetrate deep into the membrane; when increasing σ_0, it slightly increased P_0. In distinction to cortisol, the fold amplitude A increased and the fold wavelength λ decreased (Figures 1 and 2). A more precise model is needed to explain this difference.

Adrenaline has adrenoreceptors on the membrane surface, which are bound to the contractive network. Adrenaline produced significant contractions of the network, which strongly increased normal compressive stresses σ_0 in membrane and tensile stresses P_0. As a result, the fold amplitude A sharply increased. Smaller folds with the wavelength 50 nm and amplitude 2.5 nm appeared on the surface of folds with the wavelength 250 nm and amplitude 15 nm.

Testosterone exerted an effect similar to that of adrenaline. Normal compressive stresses σ_0 in the membrane and tensile stresses P_0 of the protein network increased considerably. Smaller folds with the wavelength 50 nm and amplitude 5 nm appeared on the surface of folds with the wavelength 400 nm and amplitude 10-12 nm.

Androsterone increased normal compressive stresses σ_0 in the membrane to a smaller extent. Therewith, it slightly increased compression of the protein contractive network. As a result, the fold amplitude and wavelength λ showed a minor increase.

DHEA and *DHEAS* are the most hydrophilic hormones among the hormones considered in this work. They cannot penetrate deep into the membrane and produce very weak conformational changes in membrane proteins. DHEA increased normal compressive stresses σ in the membrane as well as the wavelength λ and the amplitude of membrane folds. It did not change tension of the protein contractive network, P_0 also remained unchanged. DHEAS, which is even a more hydrophilic hormone, produced virtually no changes in the membrane conformation.

The mechanism underlying the coherence between membrane bending ζ and pressure P_B remains unclear for the case when stresses P_c created by the protein contractive network are applied to certain points on the internal surface of erythrocyte membrane. These points form a network of equiangular triangles with the side length of 100 nm. Meanwhile, the wavelength of the folds formed on the membrane could significantly differ from 100 nm (Table 2). To explain this effect, a more precise theoretical model is needed.

DISCUSSION

Hormones (adrenaline, noradrenaline, cortisol, testosterone, androsterone, DHEA and DHEAS) change the conformation of plasmatic membranes upon interaction with them (Sections 2.1 – 2.4). This effect was demonstrated for erythrocyte membranes. This is a manifestation of the non-genomic effect of hormones. Since all biomembranes have a similar structure, the conclusion made for erythrocyte membranes will hold true for other biomembranes. Hormones adsorbed on biomembranes facilitate the formation of protein-lipid domains that affect the functions of membranes. It is necessary to verify this hypothesis in the future. The effect of hormones on structural transitions in biomembranes and the effect of structural transitions on the functions of membranes and cells should be studied on membranes of different cells, particularly the stem cells, and then the mechanisms of action could be generalized.

Structural transitions in biomembranes affect the activity of Na^+, K^+-ATPases in erythrocyte membranes (Section 2.5). The effect is exerted via changing the field of mechanical stresses in biomembrane (Section 2.6). Among the driving forces of conformational transformations of the enzyme are mechanical vibrations (phonons), which are transmitted to subunits from the lipid bilayer. Although being not a single driving force of conformational transformations of the enzyme, they affect the enzyme activity. As microviscosity (stiffness) of the membrane is increased, the phonon energy of the bilayer also increases. This can explain the primary growth of the enzyme activity. As stiffness is further increased, the enzyme activity decreases (the dome-like dependence on the concentration of hormones in suspension).

When lipoprotein precursors detach from the membrane, the compression-tension zones are formed. A further increase in compression and tensile stresses can produce the structural transition gel phase $L_{\beta'} \rightarrow$

liquid crystal phase L_α in the tensile zone between lipoprotein precursor and membrane. In local tensile zones, the bridge thickness decreased, which led to necking. Density of the lipid bilayer decreased in the process, which resulted in the formation of pores and then cracks, and the lipoprotein precursor detached from the membrane.

It is known that mechanical strength of surface metal nanolayers depends on their structure. If local structural transitions cannot occur in the zones of stress microconcentrators, elastic loading of materials reaches the level of stress macroconcentrators, thus initiating the development of numerous cracks and brittle failure of materials [53, 54, 55]. The smaller are the deformation regions, the greater is the volume of material that is simultaneously involved in a plastic flow, and the higher are the strength characteristics of the loaded material. A more uniform distribution of stresses created in a sample prevents the emergence of stress macroconcentrators where the main crack is generated [56]. In plasmatic membranes, the presence of protein-lipid domains increases the threshold of mechanical stresses at which the membrane is destroyed. If ligands adsorbed on biomembranes do not initiate the formation of small protein-lipid domains (cortisol and metal oxide nanocrystals), cracks are formed in the membrane [22, 24, 29, 30]. On the contrary, if ligands interacting with biomembranes facilitate the formation of protein-lipid domains (adrenaline, noradrenaline, testosterone, androsterone), cracks do not form in the membrane [22, 24].

It is reasonable to consider a cell as a physical system whose structural changes affect its functions; however, this very efficient approach is rarely used in modern studies. In the majority of studies, a cell is considered as a platform for biochemical reactions and as a framework of a building in which the cell's life proceeds. It is believed that the time required for biomembrane renewal is much longer than the characteristic times of biochemical reactions in a cell, so the reactions are not affected by the structure of biomembranes.

Our study has demonstrated that the cell itself as a physical system can influence the occurrence of biochemical reactions by, for example, changing the field of mechanical stresses in biomembrane. Under normal physiological conditions in the organism, erythrocyte membrane undergoes a structural transition. A slight change in external factors, for example, in the concentration of noradrenaline, produces a strong change in the conformation of biomembrane. A change in the conformation of membrane proteins (Figure 3) changes the conformation of the lipid bilayer and its microviscosity (Figure 4). This structural transition in biomembrane affects the activity of their Na^+, K^+-ATPases (Figure 5). Erythrocytes become rigid and tend to slugging, which hinders their motion in microcapillaries. This may result in occlusion of microcapillaries. Presumably, the structural transition and changes of mechanical stresses in biomembrane can affect also its other functions, for example, the oxygen permeability of membrane.

CONCLUSION

The study deals with a topical subject – the effect of structural transitions in biomembranes on their properties. The effect of noradrenaline on the structure of erythrocyte membranes and activity of Na^+, K^+-ATPases in erythrocyte membranes was investigated. Upon interaction with plasmatic membranes, the hormone changes their conformation and increases their microviscosity. The surface morphology of membrane changes and folds appear on its surface. The dependence of activity of Na^+, K^+-ATPases in erythrocyte membranes on the hormone concentration has the dome-like form. At low concentrations of the hormone in suspension, the activity of Na^+, K^+-ATPases in erythrocyte membranes increases, while at high concentrations it decreases. It was shown that this effect is exerted via changing the structure of the lipid bilayer and mechanical stresses in biomembrane. The study was carried out using atomic force microscopy of erythrocyte surface and methods for measuring the intrinsic fluorescence quenching of membrane proteins and microviscosity of the lipid bilayer.

REFERENCES

[1] Ogneva, I. V. (2015). Early development under microgravity conditions. *Biophysics, 60(5),* 849–858.

[2] Dennerll, T. J.; Joshi, H. C.; Steel, V. L.; Buxbaum, R. E.; Heidemann S. R. (1988). Tension and compression in the cytoskeleton of PC-12 neurites. II: Quantitative measurements. *J. Cell Biol., 107,* 665–674.

[3] Putnam, A. J.; Schultz, K.; Mooney, D. J. (2001). Control of microtubule assembly by extracellular matrix and externally applied strain. *Am. J. Physiol. Cell Physiol., 280,* 556–564.

[4] Sukharev, S.; Corey, D. P. (2004). Mechanosensitive channels: multiplicity of families and gating paradigms. *Sci. STKE, 219,* re 4.

[5] Maroto, R.; Raso, A. T.; Wood, G.; Kurosky, A.; Martinac, B.; Hamill, O. P. (2005). TRPC1 forms the stretch-activated cation channel in vertebrate cells. *Nat. Cell Biol., 7,* 179–185.

[6] Molnár, T.; Yarishkin, O.; Iuso, A.; Barabas, A.; Jones, B.; Marc, R. E.; Phuong, T. T.; Križaj D. (2016). Store-Operated Calcium Entry in Müller Glia Is Controlled by Synergistic Activation of TRPC and Orai Channels. *J. Neurosci., 36(11),* 3184 - 3198.

[7] Lewin's Cells. (2011). Editors: Lynne Cassimeris, Vishwanath R. Lingappa, George Plopper; Jones & Bartlett Learning.

[8] Yildirmis, S.; Alver, A.; Yandi, Y. E.; Demir, S.; Senturk, A.; Bodur, A.; Mentese A. (2016). The effect of erythrocyte membranes from diabetic and hypercholesterolemic individuals on human carbonic anhydrase II activity. *Arch. Physiol. Biochem., 122(1),* 14-18.

[9] Negulyaev, Y. A.; Vedernikova, E. A.; Maximov, A. V. (1996). Disruption of actin filaments increases the activity of sodium-conducting channels in human myeloid leukemia cells. *Mol. Biol. Cell, 7.* 1857–1864.

[10] Vedernikova, E. A.; Maksimov, A. V.; Neguliaev, Iu. A. (1997). Functional properties and cytoskeletal-dependent regulation of sodium channels in leukemia cell membranes. *Tsitologiia, 39,* 1142–1151.

[11] Suzuki, M.; Miyazaki, K.; Ikeda, M.; Kawaguchi, Y.; Sakai O. (1993). F-actin network may regulate a Cl-channel in renal proximal tubule cells. *J. Membr. Biol., 134*, 31–39.

[12] Schwiebert, E. M.; Mills, J. W.; Stanton, B. A. (1994). Actin-based cytoskeleton regulates a chloride channel and cell volume in a renal cortical collecting duct cell line. *J. Biol. Chem., 269*, 7081–7089.

[13] Devarajan, P.; Scaramuzzino, D. A.; Morrow, J. S. (1994). Ankyrin binds to two distinct cytoplasmic domains of Na^+,K^+-ATPase alpha subunit. *PNAS, 91(8)*, 2965–2969.

[14] Srinivasan, Y.; Elmer, L.; Davis, J.; Bennett, V.; Angelides, K. (1988). Ankyrin and spectrin associate with voltage-dependent sodium channels in brain. *Nature, 333*, 177–180.

[15] Benos, D. J.; Awayda, M. S.; Ismailov, I. I.; Johnson, J. P. (1995). Structure and function of amiloride-sensitive Na^+ channels. *J. Membr. Biol., 143*, 1–18.

[16] Maniotis, A. J.; Chen, C. S.; Ingber D. E. (1997). Demonstration of mechanical connections between integrins, cytoskeletal filaments, and nucleoplasm that stabilize nuclear structure. *PNAS, 94*, 849–854.

[17] Huang, H.; Kamm, R. D.; Lee, R. T. (2004). Cell mechanics and mechanotransduction: pathways, probes, and physiology. *Am. J. Physiol. Cell Physiol., 287*, 1–11.

[18] Odde, D. J.; Ma, L.; Briggs, A. H.; DeMarco, A.; Kirschner, M. W. (1999). Microtubule bending and breaking in living fibroblast cells. *J. Cell. Sci., 112*, 3283–3288.

[19] Huesties, W. H.; McConnel, H. M. (1974). A functional acetylcholine receptor in the human erythrocyte. *Biochem. Biophys. Res. Commun., 57*, 732-762.

[20] Mitre-Aguilar, I. B.; Cabrera-Quintero, A. J.; Zentella-Dehesa A. (2015). Genomic and non-genomic effects of glucocorticoids: implications for breast cancer. *Int. J. Clin. Exp. Pathol., 8(1)*, 1–10.

[21] Farach-Carson, M. C.; Davis, P. J. (2003). Steroid Hormone Interactions with Target Cells: Cross Talk between Membrane and Nuclear Pathways. *J. Pharmacol. Exp. Ther., 307*, 839–845.

[22] Panin, L. E.; Mokrushnikov, P. V.; Kunitsyn, V. G.; Zaitsev B. N. (2010). The interaction mechanism of cortisol and catecholamines with structural components of erythrocyte membranes. *Journal of Physical Chemistry B, 114*, 9462-9473.

[23] Panin, L. E.; Mokrushnikov, P. V.; Kunitsyn, V. G.; Panin, V. E.; Zaitsev B. N. (2011). Fundamentals of multilevel mesomechanics of nanostructural transitions in erythrocyte membranes and their destructions in interaction with stress hormones. *Physical Mesomechanics,14(3-4),* 167-177.

[24] Panin, L. E.; Mokrushnikov, P. V.; Kunitsyn, V. G.; Zaitsev B. N. (2011). Interaction mechanism of anabolic steroid hormones with structural components of erythrocyte membranes. *Journal of physical chemistry B, 115,* 14969-14979.

[25] Mokrushnikov, P. V.; Dudarev, A. N.; Tkachenko, T. A.; Gorodetskaya, A. Y.; Usynin, I. F. (2017). Effects of native and oxidized apolipoprotein A-I on lipid bilayer microviscosity of erythrocyte plasma membrane. *Journal of Biochemistry (Moscow), Supplement Series A: Membrane and Cell Biology, 11(1),* 48-53.

[26] Panin, L. E.; Mokrushnikov, P. V. (2014). The Action of Androgens on Na^+,K^+-ATPase Activity of Erythrocyte Membranes. *Biophysics, 59(1),* 127–133.

[27] Panin, L. E.; Mokrushnikov, P. V.; Knjazev, R. A.; Kolpakov, A. R.; Zajitsev, B. N. (2012). Hormones of stress and coronary syndrome X (experimental research). *Atherosclerosis, 8,* 5-10.

[28] Mokrushnikov, P. V.; Panin, L. E.; Zaitsev B. N. (2015). The action of stress hormones on the structure and function of erythrocyte membrane. *Gen. Physiol. Biophys., 34(3),* 311-321.

[29] Mokrushnikov, P. V.; Panin, L. E.; Doronin, N. S.; Zaitsev, B. N.; Kozelskaya, A. I.; Panin, A. V. (2011). Interaction of corundum and quartz nanocrystals with erythrocyte membranes. *Biophysics, 56(6),* 1074–1077.

[30] Kozelskaya, A. I.; Panin, A. V.; Khlusov, I.A.; Mokrushnikov, P. V.; Zaitsev, B. N.; Kuzmenko, D. I.; Vasyukov G. Yu. (2016).

Morphological changes of the red blood cells treated with metal oxide nanoparticles. *Toxicology in Vitro, 37,* 34–40.
[31] Dawson Rex, M. C.; Elliot, D. C.; Elliot, W. H.; Jones, K. M. Data for biochemical research. Oxford: Clarendon Press; 1986.
[32] Attalah, N. A.; Lata, G. F. (1968). Steroid-protein interactions studied by fluorescence quenching. *Biochemical et biophysical acta, 168,* 321–333.
[33] Ooi, T.; Itsuka, E.; Onari, S. Biopolymers. Ed. by Y. Imanishi. Tokyo: Kyoritsu Shuppan; 1985.
[34] Simons, K.; Ikonen, E. (1997) Functional rafts in cell membranes. *Nature, 387,* 569 – 571.
[35] Simons, K.; Gerl, M. J. (2010) Revitalizing membrane rafts: new tools and insights. *Nat. Rev. Mol. Cell Biol, 11,* 688–699.
[36] Panin, L. E.; Mokrushnikov, P. V.; Panin, A. V.; Khokhlova, A. I.; Shugurov, A. R. (2012) Wrinkling-induced fracture of biological membranes under stress. 19th European Conference on Fracture (ECF19), Kazan, Russia.
[37] Kozelskaya, A. I.; Panin, A. V.; Panin, L. E.; Mokrushnikov, P. V.; Kuzmenko, D. I.; Chernikov, A. V. (2013) Regularities in elastic corrugation of erythrocyte membranes. Hierarchically organized systems of living and inanimate nature. Proc. International Conference. Tomsk, 9-13 Sept. 2013 (in Russian).
[38] Kazennov, A. M.; Maslova, M. N. (1991) The effect of the membrane skeleton of non-nucleated erythrocytes on the properties of transport ATPases. *Tsitologiya 33,* 32–41 (in Russian).
[39] Arzamazova, N. M.; Aristarkhova, E. A.; Shafieva, G. I.; Nazimov, I. V.; Aldanova, N. A.; Modyanov, N. N. (1985) Primary structure of Na^+,K^+-ATPase α-subunit. I. Analysis of hydrophilic fragments of polypeptide chain. *Bioorg. Chemistry,* 11 (12), 1598-1606 (in Russian).
[40] Petrukhin, K. E.; Broude, N. E.; Arsenyan, S.G.; Grishin, A. V.; Dzhandzhugazyan, K. E.; Modyanov, N.N. (1985) Primary structure of Na,K-ATPase alpha-subunit. II. Isolation, reverse transcription and

cloning of matrix RNA. *Bioorg. Chemistry*, 11(12), 1607–1613 (in Russian).

[41] Broude, E. E.; Monastyrskaya, G. S.; Petrukhin, K. E.; Grishin, A. R.; Kiyatkin, E. I.; Melkov, A. M.; Smirnov, E. R.; Sverdlov, E. E.; Malyshev, E. E.; Modyanov, N. N. (1987) Primary structure of Na,K-ATPase beta-subunit of pig kidneys. Reverse transcription, cloning of mRNA, and identification of complete nucleotide sequence of the structural part of the gene. *Bioorgan. Chemistry*, 1 (13), 14–19 (in Russian).

[42] Bouvraisa, H. F.; Cornelius; Ipsena, John H.; Mouritsena Ole G. (2012) Intrinsic reaction-cycle time scale of Na^+,K^+-ATPase manifests itself in the lipid–protein interactions of nonequilibrium membranes. *PNAS, 109, 45*, 18442–18446.

[43] Rowlands, S.; Sewchand, L. S.; Enns E. G. (1982) A quantum mechanical interaction of human erythrocytes. *Can. J. Physiol. Pharmacol.*, 60, 52 - 59.

[44] Rheinstadter, M. C.; Schmalzl, K.; Wood, K.; Strauch D. (2009) Protein-protein interaction in purple membrane. *Phys. Rev. Lett.*, 103, 128104.

[45] Mason, W. P. Lattice Dynamics. New York and London: Academic Press; 1965.

[46] Fanton, L.; Belhani, D.; Vaillant, F.; Tabib, A. (2009) Heart lesions associated with anabolic steroid abuse: Comparison of post-mortem findings in athletes and norethandrolone- induced lesions in rabbits. *Exp. Toxicol Pathol., 61(4)*, 317–323.

[47] Liu, S. C.; Derick, L. H.; Palek J. (1987) Visualization of the hexagonal lattice in the erythrocyte membrane skeleton. *Journal of Cell Biology, 104*, 527 – 536.

[48] Vladimirov, Yu. A.; Dobretsov G. E. Fluorescent Probes in the Study of Biological Membranes. Moscow: Nauka; 1980.

[49] Landau, L. D.; Lifshits, E. M. Theoretical Physics. 10 volumes. Vol. VII. Theory of Elasticity. Moscow: Nauka; 1987 (in Russian).

[50] Palek, J.; Sahr, K. E. (1992) Mutations of the red blood cell membrane proteins: from clinical evaluation of the underlying genetic defects. *Blood,* 80 (2), 308-320.

[51] Tolpigo, V. K.; Clarke D. R. (1998) Wrinkling of α – alumina films grown by thermal oxidation - I. Quantitative stadies on single crystals of Fe-Cr-Al alloy. *Acta mater.,* 46, 5151 - 5166.

[52] Tolpigo, V. K.; Clarke D. R. (1998) Wrinkling of α – alumina films grown by thermal oxidation - II. Oxide separation and failure. *Acta mater.,* 46, 5167 - 5174.

[53] Panin, V. E.; Egorushkin, V. E. (2008) Nonequilibrium thermodynamics of a deformed solid as a multiscale system. Corpuscular-wave dualism of plastic shear. *Physical Mesomechanics, 11 (3–4),* 105-123.

[54] Panin, V. E.; Egorushkin, V. E. (2010) Nanostructural states in solids. *The Physics of Metals and Metallography, 110(5),* 464-473.

[55] Panin, V. E.; Egorushkin, V. E. (2011) Deformable solid as a nonlinear hierarchically organized system. *Physical Mesomechanics, 14 (5-6),* 207-223.

[56] Panin, V. E.; Sergeev, V. P.; Panin, A. V.; Pochivalov, Yu. I. (2007) Nanostructuring of surface layers and production of nanostructured coatings as an effective method of strengthening modern structural and tool materials. *The Physics of Metals and Metallography, 104 (6),* 627-636.

In: Lipid Bilayers
Editor: Mohammad Ashrafuzzaman

ISBN: 978-1-53616-392-6
© 2019 Nova Science Publishers, Inc.

Chapter 3

THE ROLE OF THE LIPID BILAYER IN THE ERYTHROCYTE MEMBRANE STRUCTURAL CHANGES

Ivana Pajic-Lijakovic[*] *and Milan Milivojevic*
Faculty of Technology and Metallurgy, Belgrade University,
Belgrade, Serbia

ABSTRACT

In this chapter, an attempt was made to discuss and connect various modeling approaches on various time and space scales in order to shed further light on the lipid bilayer role in the erythrocyte membrane structural changes under isotonic and hypotonic conditions.

Local changes of the bilayer bending state enhance anomalous sub-diffusion and eventually lead to hop-diffusion of lipids. These effects induce the anomalous nature of energy dissipation during lipids structural ordering and could have feedback impact on the bilayer bending state. The bilayer structural changes influence the positive hydrophobic mismatch effects between trans-membrane protein band 3 and lipids. These effects could lead to the protein tilt angle changes and the protein clustering. Tilt

---

[*] Corresponding Author's Email: iva@tmf.bg.ac.rs.

angle changes influence packing state of band 3 clusters and their association-dissociation to spectrin. The bilayer coupling with the actin-spectrin cortex influences: (1) conformational changes of spectrin filaments and (2) the bilayer bending state. Cumulative effects of these changes cause anomalous nature of the erythrocyte membrane viscoelasticity. The spectrin filament conformations are dependent on the state of three types of the complexes: (1) band 3 complexes with ankyrin located at the midpoint of the filament, (2) band 3 complexes with adducin located at the spectrin-actin junction complexes, and (3) band 3 complexes with spectrin located along the spectrin filaments. Spatial distribution of band 3 molecules and their clustering are influenced by bending state of the lipid bilayer. Cause-consequence relations between main membrane constituents are discussed based on thermodynamical and rheological modeling approaches.

Keywords: the rearrangement of lipids, the lipid bilayer bending state, the bilayer-cortex coupling, positive hydrophobic mismatch effects

1. INTRODUCTION

Lipid bilayer significantly influences the rheological behavior of the erythrocyte membrane through interactions with the actin-spectrin cortex and the trans-membrane protein band 3. Lipids do not mix ideally and form a gel and fluid phases or two fluid phases depending on external conditions, i.e., temperature, tonicity (i.e., the ionic strength of external solution) (Almeida, 2009). Consequently, their composition is not uniform within the erythrocyte lipid bilayer. Lipids are segregated within cholesterol-enriched microdomains called "rafts" (Simons and Toomre, 2000; Mikhalyov and Samsonov, 2011). These rafts consist of smaller compartments. Lateral diffusion of lipids within these compartments has been 5-6 times higher than the macroscopic diffusion due to the cumulative resistance effects of molecules hopping (Fujiwara et al., 2002; 2016). Lipids are confined within a compartment for about 11 ms before hopping to an adjacent compartment. Spatial distribution of lipid molecules and their conformations depend on the bilayer bending state (Leonard et al., 2017). The bending state could be influenced by mechanical or osmotic stresses (Pajic-Lijakovic and Milivojevic, 2014; Pajic-Lijakovic 2015a).

Cholesterol molecules stabilize changes in the lipid bilayer bending state (Choubey et al., 2013; Lundbaek et al., 2010; Leonard et al., 2017). Leonard et al. (2017) reported that cholesterol-enriched domains gathered in high curvature area caused by cell deformation. Consequently, cholesterol leads to a decrease in bending free energy. In contrast, sphingomyelin-enriched domains increased in abundance upon calcium efflux during shape restoration.

The lateral diffusion coefficient of cholesterol decreases with the increase of cholesterol local concentration. Trans-membrane protein band 3 also reduced the mobility of cholesterol (Golan et al., 1984). Consequently, the spatial distribution of band 3 molecules influences the bending state of the lipid bilayer. At 37 °C, the diffusion coefficient of the lipid probes are equal to $D_{Lmacro} = 2.1 x 10^{-13} \frac{m^2}{s}$, or approximately 4 times greater than the fastest diffusion of the major erythrocyte trans-membrane protein, band 3, in the same membrane (Golan and Veatch, 1980).

However, lipid rearrangement near the inclusions such as the band 3 molecules has the feed-back impact on the band 3 molecules conformation state and their spatial distribution. Band 3 molecules form clusters caused by positive hydrophobic mismatch effects during their lateral diffusion through lipid surrounding (De Meyer et al., 2010; Standberg et al., 2012; Milovanovic et al., 2015). These effects induce the rearrangement of lipids near the inclusions and also could induce the protein tilting (Pajic-Lijakovic and Milivojevic, 2017) as was shown in Figure 1.

Clustering of integral proteins such as band 3 could be treated as the first-order phase transition (Gil et al., 1998; Sens and Tarner, 2004). This phase transition through homogeneous nucleation is induced by: (1) short-range attractive protein-protein interactions and (2) long-range lipid-mediated repulsive interactions between protein clusters (Sens and Tarner, 2004; Destainville, 2008).

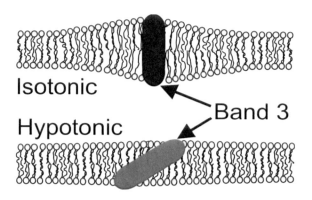

Figure 1. Positive hydrophobic mismatch effects between band 3 molecules and the lipid bilayer causes the protein tilting.

This cause-consequence relation is significantly influenced by the bending of the bilayer under hypotonic conditions. The membrane response includes the successive sub-bioprocesses: (1) erythrocyte swelling, (2) lifetime of the lipid structural integrity and the rearrangements of trans-membrane protein band 3, (3) the reversible hemolytic hole formation and hemoglobin release to surrounding solution (Seeman et al., 1973; Sato et al., 1993; Zade-Oppen, 1998; Pribish et al., 2002; Pajic-Lijakovic, 2015a). These changes induce anomalous nature of energy dissipation during (1) lipids structural organization and (2) the cortex rearrangement and could be quantified by the effective moduli (Pajic-Lijakovic and Milivojevic, 2014). The protein tilting caused by hydrophobic mismatch effects under hypotonic conditions is the prerequisite for hemolytic hole formation during band 3 cluster excitation (Leiber and Steck, 1982; Sato et al., 1993; Pajic-Lijakovic et al., 2010). The hemolytic hole represents the change of the cluster packing state from close packing to ring-like structure caused by fluctuations of the lipid bilayer after erythrocyte swelling. Achieving of the ordered, ring-like structure of the excited cluster instead of cluster disintegration, is primarily caused by protein tilting effects (De Meyer et al., 2010; Strandberg et al., 2012; Milovanovic et al., 2015; Pajic-Lijakovic and Milivojevic, 2017). Protein tilting is caused by long-range the lipid curvature mediated interactions and short-range lipid compression (Argudo et al., 2016). To our opinion, the tilting represents the key factor which could ensure the integrity of the membrane under hypotonic conditions.

Local changes of the lipid bilayer bending state influence the structural changes of the actin-spectrin cortex: (1) directly through the bilayer-cortex coupling (Pajic-Lijakovic and Milivojevic, 2014) and (2) indirectly by influencing the spatial distribution of band 3 molecules (Pajic-Lijakovic, 2015a; 2015b). Spatial distribution of band 3 molecules has an impact on the conformation of the spectrin filaments as well as their flexibility. The spectrin filament's conformations are dependent on the state of three types of complexes: (1) band 3 complexes with ankyrin located at the midpoint of the filament, (2) band 3 complexes with adducin located at the spectrin-actin junction complexes and (3) band 3 complexes with spectrin located along the spectrin filaments (Golan and Vetch, 1980; Gov, 2007; Franco and Low, 2010; Kodippili et al., 2012). Complex dynamics of the single filament conformations is connected with the cumulative effects of protein's complexes changes. The spectrin flexibility varies from purely flexible to semi-flexible (Gov and Safran, 2005; Le et al., 2005). It is influenced by (1) the number of band 3 molecules attached to single spectrin filaments, and (2) phosphorylation of the actin-junctions. The flexibility distribution of the spectrin leads to the formation of the cortex microdomains (Pajic-Lijakovic and Milivojevic, 2014; 2015b). The size of cortex microdomains influences the cortex-bilayer coupling and consequently the viscoelasticity of the whole membrane.

The aim of this theoretical consideration is to point out the cause-consequence relations between the main membrane constituents: (1) the actin cortex, (2) the lipid bilayer, and (3) the trans-membrane protein band 3 under isotonic and hypotonic conditions obtained at various time and space scales. For deeper insight into the membrane viscoelasticity, it is necessary to estimate: (1) interactions between single band 3 molecules and surrounding lipids (Lundbaek et al., 2010), (2) the cumulative effects of band 3 rearrangement and their influence to the bilayer bending state and the conformational changes of spectrin filaments (Gov and Safran, 2005; Li et al., 2005; Pajic-Lijakovic, 2015b), and (3) the cortex-bilayer coupling under isotonic and hypotonic conditions. The cortex-bilayer coupling influences the collective phenomena such as lipid phase separation and the formation of lipid microdomains and cortex microdomains (Elson et al., 2010; Pajic-Lijakovic and Milivojevic, 2014; Pajic-Lijakovic, 2015a). These complex phenomena influence energy storage and energy dissipation during the membrane structural changes under various microenvironmental conditions. Energy storage and dissipation represent the product of the membrane viscoelasticity. The membrane structural changes were discussed based on thermodynamical and rheological modeling considerations at the time scales from milliseconds to tens of minutes.

2. THE ERYTHROCYTE MEMBRANE UNDER ISOTONIC AND HYPOTONIC CONDITIONS

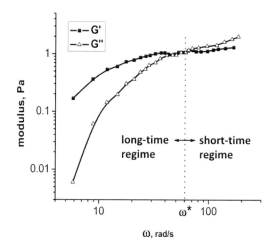

Figure 2. The erythrocyte membrane viscoelasticity-two regimes.

The membrane thermal fluctuations under isotonic conditions induce alternating expansion and compression of the membrane parts in order to ensure surface and volume conservation. The membrane structural changes were considered within two-time regimes: (1) long-time regime (regime 1) for $t \in [0.20, 1.05\ s]$ and (2) short-time regime (regime 2) for $t \in [0.03,\ 0.20\ s]$ (where t is the membrane relaxation time) (Pajic-Lijakovic and Milivojevic, 2014). The regimes were shown schematically in the context of storage and loss moduli vs. angular velocity in Figure 2. Storage modulus represents the measure of energy storage while loss modulus represents the measure of energy dissipation during the membrane structural changes. Storage and loss moduli increase faster in long-time regime 1 relative to short-time regime 2.

Spectrin inter-chain interactions, the short-time bending relaxation of the lipid bilayer, and the short-time motion of band 3 molecules are in millisecond scale – regime 2 (Tomishige et al., 1998; Gov 2007, Pajic-Lijakovic and Milivojevic, 2014; 2015b). Short-time bending relaxation is the necessary time for cholesterol diffusion within the curvature (Elson et al., 2010). Spectrin inter-chain interactions correspond to the conformational changes of the spectrin parts within the blobs between two neighbor mid-point attachments made by band 3 molecules as was shown in Figure 3.

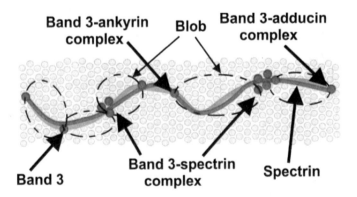

Figure 3. The configuration changes of spectrin filament influenced by bending of the lipid bilayer and the attachment of band 3 molecules.

Regime 1 accounts for the cumulative effects of (1) the long-time changes of the bilayer bending states, (2) the spectrin intra filaments structural changes which induce the ordering of spectrin and actin within the cortex microdomains, and (3) the long-time spatial rearrangements of band 3.

The membrane structural changes under hypotonic conditions are much complex than that under isotonic condition due to the cell volume increase while the membrane surface remains approximately constant. These changes occur within the successive multi-time sub-processes (Pajic-Lijakovic, 2015a): (1) erythrocyte swelling, (2) lifetime of the lipid structural integrity and the rearrangements of trans-membrane protein band 3, (3) the reversible hemolytic hole formation, and (4) hemoglobin (Hb) release to surrounding solution. The term "membrane lifetime" is introduced by Pribish et al. (2002). It represents the time period after erythrocyte swelling and before the formation of hemolytic hole. Two processes: (1) the membrane relaxation to swelling and (2) the integrity of the membrane preserving process lead to long-time rearrangement of the membrane constituents. The membrane "lifetime" is near to duration of hypotonic hemolysis and depends on the erythrocyte membrane viscoelasticity.

The membrane relaxation time under hypotonic conditions represents the necessary time for the membrane structural adaptation to new environmental conditions. Duration of the membrane relaxation per single cell t depends on contributions of three sub-processes: (1) time for cell swelling $t_{sw} \epsilon [5, 100\ s]$, (2) membrane lifetime $\tau_m \epsilon (0, t_H)$ (where t_H is the hemolytic time), and (3) time for Hb release from already formed hemolytic hole during successive open-closed state changes $t_{release\ eq} = n_{jumps}(t_{o\ eq} + t_{c\ eq})$ (where $n_{jumps} = 1 - 8$ is the number of repeated open/closed cycles, $t_{o\ eq} \approx 0.27\ s$ is the hole opening time period, $t_{c\ eq} \approx 0.26\ s$ is the hole closing time period) such that $t_{release\ eq} \in [1,5\ s]$ (Zade-Oppen, 1998; Pribush et al., 2002). The time $t_{c\ eq}$ is the necessary time for the membrane repeated stretching as the consequence of solvent repeated in-flow driven by a total osmotic pressure difference between the intracellular region and external medium. Zade-Oppen (1998) observed repetitive erythrocyte "jumps" during Hb release after erythrocyte swelling. Every jump corresponds to the repeated hemolytic hole opening period $t_{o\ eq}$. The relaxation time for cell swelling under hypotonic conditions depends on the solution tonicity. Pribush et al. (2002) reported that the swelling time is equal to 15 s for the corresponding external solution made by isotonic solution and water in the ratio 1:5. When the solution tonicity increases up to the ratio 1:1.5 the swelling time increases up to 75 s. The rearrangement times of band 3 molecules under hypotonic conditions are: ~500 s under tonicity of 5.2 mM NaPO$_4$ at 21°C and ~1200 s under tonicity of 5.2 mM NaPO$_4$ at 21°C (Golan and Veatch, 1980). Consequently, the membrane relaxation time under hypotonic conditions is between (1) long-time limit $t \sim \tau_m$, and (2) short-time limit $t \sim t_{sw}$ depending on the membrane viscoelasticity and experimental conditions (Pajic-Lijakovic, 2015a).

2.1. The Lipid Bilayer under Isotonic and Hypotonic Conditions

Local changes of the bilayer bending state are more intensive under hypotonic conditions than under isotonic condition. The thickness of the bilayer decreases during erythrocyte swelling under hypotonic conditions (Asami and Yamaguchi, 1999). These changes enhance anomalous sub-diffusion and eventually lead to hop-diffusion of lipids (Trimbe and Grinstein, 2015). The bilayer structural changes have the feedback impact to band 3 protein-lipids positive hydrophobic mismatch effects (De Meyer et al., 2010; Standberg et al., 2012; Milovanovic et al., 2015). These long-range interactions are caused by local curvature changing while short-range interactions are caused by local compressing of the lipid bilayer (Lundbaek et al., 2010Argudo et al., 2016).

De Meyer et al. (2010) pointed to the cholesterol role in lipid-mediated protein-protein interactions. These effects could lead to the protein tilt angle changes and the protein clustering. Self-association tendency between the highly anionic N-terminal domains of band 3 molecules is caused by short-range attractive protein-protein interactions (Taylor et al., 1999). Tilt angle changes influence the packing state of band 3 clusters and their association-dissociation to spectrin (Golan and Veatch, 1980; Blackman et al., 1996). Band 3 molecules form various complexes with spectrin and influence its conformational changes. Detail description of these complexes will be given in the next section.

2.2. Band 3 Molecules under Isotonic and Hypotonic Conditions

The total number of band 3 molecules per erythrocyte surface is $n_T \approx 1 \times 10^6$ (Saxton, 1990). They could form three types of complexes (Pajic-Lijakovic, 2015b). Schematic representation of these three types of band 3 complexes is shown in Figure 4.

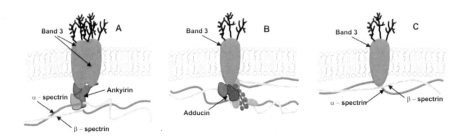

Figure 4. Band 3 complexes with: (a) ankyrin, (b) adducin, and (c) spectrin.

The first type (20-40%) are high-affinity complexes with ankyrin (quantified by the dissociation constant ~5 nM) (Tomishige et al., 1998; Kodippili et al., 2012). Band 3-ankyrin complexes are located near the center of spectrin tetramers. These complexes could survive hypotonic conditions. Golan and Vetch (1980) reported that 29% of band 3 population remains attached to the cortex under ionic strength 26 mM $NaPO_4$ solution at 30°C. The second type (~30%) forms lower affinity complexes with the adducin (quantified by the dissociation constant ~100 nM) (Franco and Low, 2010; Kodippili et al. 2012). These complexes are located at the spectrin-actin junctions. The junctions of the network link 4-7 spectrin filaments. The rest of the band 3 molecules forms low-affinity complexes with spectrin filaments. The affinity of these complexes is quantified by the dissociation constant ~1-10 μM (Golan and Veatch, 1980). All band 3 molecules through these complexes contribute to the spectrin conformational changes by reducing its mobility and induce the cortex stiffening (Pajic-Lijakovic, 2015b).

Lower affinity complexes with adducing and with spectrin could be dissociated under hypotonic conditions. Golen and Veatch (1980) determined that equilibrium volume fraction of freely diffuse band 3 molecules at 21°C is: (1) 11 ± 9 % under the high ionic strength of external medium (46.0 mM NaPO$_4$) and (2) 72 ± 7 % under the low ionic strength of external medium (5.2 mM NaPO$_4$).

2.3. The Spectrin-Actin Cortex under Isotonic and Hypotonic Conditions

Conformational changes of spectrin filaments depend on the number of attached band 3 molecules (Figure 3). The attachment of spectrin to these complexes is a transient dissociation process driven by ATP hydrolysis and phosphorylation. Parts of the spectrin filament between two mid-point attachments behave as independent blobs (Gov, 2007). The flexibility of the filament parts depends on $\frac{l_s}{L_p}$ (where l_s is the length of the filament part and L_p is the spectrin persistence length $L_p = 15 - 25\ nm$ (Boal, 2012). Three conditions are expected (Pajic-Lijakovic, 2015a; Pajic-Lijakovic and Milivojevic, 2017):

(1) if $\frac{l_s}{L_p} \gg 1$ the filament parts are flexible,

(2) if $\frac{l_s}{L_p} \approx 1$ the filament parts are semi-flexible, and

(3) if $\frac{l_s}{L_p} \ll 1$ the filament parts are rod-like (Pajic-Lijakovic, 2015b).

Average contour length of the spectrin filaments is 200 nm.

3. INTERACTIONS OF BAND 3 MOLECULES WITH THE SURROUNDING LIPID BILAYER

Firstly, we considered interactions between (1) the lipid bilayer with single band 3 molecule (the inclusion) and (2) the lipid bilayer and single spectrin filament under isotonic and hypotonic conditions. After that, we considered interactions between the lipid bilayer, spectrin filaments and band 3 molecules at the mesoscopic level.

Band 3 molecules as the rigid inclusions induce the local bending and compression of the surrounding lipid bilayer. The bilayer structural changes have the feedback impact to band 3 protein-lipids positive hydrophobic mismatch effects (De Meyer et al., 2010; Standberg et al., 2012; Milovanovic et al., 2015).

Free energy of the surrounding bilayer has been considered as the sum of (1) the local energy of bending and (2) the local energy of compression. Lundbaek et al. (2010) expressed the local free energy of the bilayer deformation as:

$$\Delta G_{def} = \int_{r_0}^{\infty} \left\{ \frac{K_a}{2} \left(\frac{d-d_0}{d_0} \right)^2 + \frac{K_c}{2} \left(\frac{c_1 - c_2}{2} - c_0 \right)^2 \right\} 2\pi r dr - \int_{r_0}^{\infty} \frac{K_c}{2} c_0^2 \, 2\pi r dr \qquad (1)$$

where ΔG_{def} is the deformation energy of the lipid bilayer, K_a is the compressive modulus, K_c is the bending modulus, d is the perturbed bilayer thickness at $r = r_0$, r_0 is the radius of inclusion, d_0 is the thickness of unperturbed bilayer, c_1 and c_2 are the curvatures of the bilayer caused by the presence of the inclusion, c_0 is the intrinsic curvature. The perturbed bilayer thickness is approximately equal to $d \approx l$ (where l is the length of the inclusion). Eq. 1 can be formulated by Lundbaek et al. (2010) as:

$$\Delta G_{def} = H_B (l - d_0)^2 + H_X (l - d_0) c_0 - H_C c_0^2 \qquad (2)$$

where H_B, H_X, and H_C are the elastic coefficients which depend on the bending modulus K_c and the compressive modulus K_a. These moduli are influenced by Coulomb interactions between inclusion and lipids as well as within the lipids themselves. Compression of surrounding lipids induces the decrease of the average lipid-lipid distance which leads to electrostatic interactions between charged lipids. These interactions can induce the local bilayer stiffening (Ashrafuzzaman and Tuszynski, 2012). The stiffening influences the bilayer bending which has the feedback impact on the state of band 3 molecules. Direct Coulomb interactions depend on protein charge and the average protein-lipid distance. Consequently, complex electrostatic phenomena cause the additional energy dissipation which could change the local rheological behavior of the bilayer.

Ashrafuzzaman and Tuszynski (2012) reported that these electrostatic phenomena between proteins and lipids also induce the conformational changes of proteins which regulate the function of the ionic channel.

The thickness of the unperturbed bilayer under the isotonic condition is equal to $d_0 = d_0^I$. However, after erythrocyte swelling under hypotonic conditions, the bilayer thickness decreases up to $d_0 = d_0^H$, such that $d_0^H < d_0^I$. The assumed value of the lipid bilayer thickness of swollen erythrocyte is 3-4 nm (Asami and Yamaguchi, 1999). This decrease of the bilayer thickness could induce an increase of the deformation energy of the lipid bilayer ΔG_{def} and the bilayer stiffening which could destabilize the bilayer structure near the inclusion. The bilayer comes into equilibrium state by minimizing the energy ΔG_{def} as:

$$\frac{\partial \Delta G_{def}}{\partial l} \to 0 \text{ for } l \to l^H \qquad (3)$$

where l^H is the length of inclusion under hypotonic conditions. The length of inclusion under hypotonic conditions l^H is lower than the length of inclusion under the isotonic condition l^I, i.e., $l^H < l^I$.

This important result points out that the stretching of the lipid bilayer during erythrocyte swelling under hypotonic conditions induces the tilting of band 3 molecules. These tilting effects of band 3 clusters caused by band 3 interactions with the surrounding lipid molecules are the prerequisite for the formation of the reversible hemolytic hole under hypotonic conditions (Pajic-Lijakovic and Milivojevic, 2017).

4. THE INFLUENCE OF THE BILAYER BENDING AND THE ATTACHMENT OF BAND 3 MOLECULES ON CONFORMATIONS OF THE SINGLE SPECTRIN FILAMENT

The spectrin-actin cortex is slightly stretched and under tension even in un-deformed state of erythrocyte due to coupling with the lipid bilayer (Gov and Safran, 2005). The end-to-end distance of the spectrin comes from equilibration of the spectrin conformational force on one side and the lipid bilayer curving force F_{curv} on the other during their coupling. Gov and Safran (2005) expressed the bilayer curving force as:

$$F_{curv} = \kappa_{eff} \frac{\langle r_g^2 \rangle}{R^3} \qquad (4)$$

where κ_{eff} is the effective bending modulus, $\langle r_g^2 \rangle^{1/2}$ is the average spectrin radius of gyration equal to $\langle r_g^2 \rangle^{1/2} \sim 70\ nm$ and $R \approx 80 - 100\ nm$ is the end-to-end distance of spectrin filaments in the spectrin-actin cortex. Effective bending modulus could be expressed as:

$$\kappa_{eff} = \kappa + \Delta\kappa(N_B) \qquad (5)$$

Where $\kappa \approx 2 \times 10^{-20}\ J$ is the average bending modulus of the lipid bilayer and $\Delta\kappa(N_B)$ is the contribution to the bending modulus due to the presence of N_B molecules of band 3 attached to the single spectrin filament.

The lipid bilayer curving force F_{curv} is equilibrated to the spectrin conformational force:

$$F_{curv} = F_s \qquad (6)$$

where F_s is the spectrin conformational force equal to $F_s = F_s(L_{p\ eff})$, while $L_{p\ eff}$ is the effective persistence length of spectrin filaments.

Effective persistence length depends on the number of attached band 3 molecules (Figure 3) which can reduce spectrin mobility and induces the increase of the effective persistence length of spectrin. Consequently, the effective persistence length has been expressed by Pajic-Lijakovic (2015b) and Pajic-Lijakovic and Milivojevic (2017) as:

$$L_{p\,eff}(T, N_B) = L_p(T) + \Delta L_p(N_B) \qquad (7)$$

Where $L_{p\,eff}(T, N_B)$ is the effective persistence length of spectrin filament, $L_p(T)$ is the spectrin persistence length for the filaments without band 3 mid-point attachments made by band 3-spectrin low-affinity complexes at the same temperature conditions and $\Delta L_p(N_B)$ is the contribution to the persistence length caused by the band 3 mid-point attachments, and T is temperature.

The spectrin filaments without the band 3 mid-point attachments satisfy the condition $\frac{L_c}{L_p} \gg 1$ and could be treated as flexible (where L_c is the contour length of the spectrin filaments). For this condition spectrin conformation force has been expressed by Gov and Safran (2005):

$$F_s \approx N\mu_s \left(R - \langle r_g^2 \rangle^{1/2} \right) \qquad (8)$$

where $N \approx 3$ is the number of spectrin filaments per network units, μ_s is the surface shear modulus of the spectrin-actin network equal to $\mu_s = \frac{k_B T}{\langle r_g^2 \rangle} \approx 6 \times 10^{-6} \frac{J}{m^2}$, k_B is Boltzmann constant.

Attached band 3 molecules on spectrin filaments caused by increasing of the effective persistence length of the spectrin. If the condition for semi-flexible filaments $\frac{L_c}{L_{p\,eff}} \approx 1$ is satisfied, the spectrin conformation can be described by worm-like force. Li et al. (2005) expressed by the worm-like force for spectrin filaments as:

$$F_s = \frac{k_B T}{L_{p\,eff}} \left\{ \frac{1}{4(1-x)^2} - \frac{1}{4} + x \right\} \qquad (9)$$

where $x = \dfrac{R - \langle r_g^2 \rangle^{1/2}}{L_c}$ is the stretch ratio and k_B is Boltzmann constant.

5. BENDING ENERGY OF THE LIPID BILAYER – THERMODYNAMICAL MODELING CONSIDERATION AT MESOSCOPIC LEVEL

Reversible structural changes of the lipid bilayer obtained under small curvature changes could be treated thermodynamically based on Helfrich type bending free energy. This theory describes the bilayer as an elastic body. The Helfrich type bending free energy functional has been modified to consider the influence of spectrin conformations and the presence of band 3 molecules on bending energy of the lipid bilayer (Shlomovitz and Gov, 2008; Kabaso et al., 2011; Pajic-Lijakovic, 2015b; Pajic-Lijakovic and Milivojevic, 2017):

$$E_B(\phi) = \int \frac{1}{2} \kappa_{eff}(\phi)(H - \phi \bar{H}_s)^2 d^2 r \qquad (10)$$

where $\phi = \phi(r,t)$ is the local volume fraction of inclusions such as band 3 molecules in the erythrocyte membrane, $\kappa_{eff}(\phi)$ is the effective bending modulus of the lipid bilayer, H is the local mean curvature, $\bar{H}_s$ is the curvature induced by conformations of the surrounding spectrin filaments. Kabaso et al. (2011) expressed the curvature $\bar{H}_s$ by accounting for the contributions of two types of spectrin filaments. Filaments of the first type are grafted at one end, or at both ends but not connected to the stretched cortex. These filaments induce a concave spontaneous curvature of radius R_{L1}. Filaments of the second type are grafted at both ends and represent a part of the connected stretched network. These filaments induce a local convex spontaneous curvature of radius R_{L2} such that $R_{L1} = -R_{L2}$. The corresponding curvature of the bilayer is equal to:

$$\bar{H}_s = n_1 \bar{H}_1 + n_2 \bar{H}_2 \tag{11}$$

where $\bar{H}_1 = \frac{1}{R_{L1}}$ and $\bar{H}_2 = \frac{1}{R_{L2}}$ are the two spontaneous curvatures associated with two types of the filaments, while n_1 and n_2 are the relative densities of the two types of spectrin filaments, normalized by saturating (maximal) packing density of the filaments n_{sat}. The effective bending modulus of the lipid bilayer is expressed by Shlomovitz and Gov (2008) as:

$$\kappa_{eff}(\phi) = \kappa(1 - \phi) + \kappa'\phi \tag{12}$$

where κ is the bending modulus of the lipid bilayer without inclusions, κ' is the contribution of inclusions to the bending modulus. Bending modulus κ represents the energy required for small deformation of the membrane from its natural curvature expressed as: $\kappa = 2\kappa_{ML} = \frac{2}{A_0}\frac{\partial^2 F_{B0}}{\partial H^2}/_{H \to 0}$ (where κ_{ML} is the bending modulus of the monolayer, A_0 is the average area for single lipids, F_{B0} is the surface free energy of the lipid bilayer without inclusions, and the condition $H \to 0$ represents the small curvature change) (Kheyfets et al., 2016).

6. THE BILAYER-CORTEX COUPLING – RHEOLOGICAL MODELLING CONSIDERATION AT MACROSCOPIC LEVEL

Microrheological experimental data for the storage and loss moduli of the erythrocyte membrane as a function of angular velocity (Amin et al., 2007; Puig-de-Morales-Marinkovic et al., 2007; Popescu et al., 2007) under isotonic condition are used to develop constitutive model equation by Pajic-Lijakovic and Milivojevic (2014). The model accounts for: (1) the viscoelasticity of the spectrin-actin cortex, (2) the viscoelasticity of the bilayer, and (3) the bilayer-cortex coupling within two-time regimes (Figure 2). Short-time regime accounts for: (1) the spectrin inter-chain interactions, (2) the diffusion of band 3 molecules, and (3) the local changes of the bilayer bending state. Long-time regime accounts for: (1) the spectrin intra-chain interactions, (2) band 3 spatial distributions, and (3) the cumulative effects of changes the bilayer bending state which lead to the formation of the cortex microdomains (Pajic-Lijakovic and Milivojevic, 2014; Pajic-Lijakovic, 2015b). The membrane stress represents the sum of two contributions: (1) the cortex stress and (2) the bilayer stress. The membrane strain is equal for the bilayer and the cortex. It is the consequence of the membrane lamellar structure. The membrane stress-strain relations for long-time regime (regime 1) and short-time regime (regime 2) are expressed as:

$$\sigma(t_R) = B\, D_t^{-(\alpha+1)}\varepsilon(t) + \eta_{effC} D_t^{\alpha}\, \varepsilon(t) + \eta_{effL} D_t^{2\alpha}\, \varepsilon(t) \text{ for regime 1}$$

(13)

$$\sigma(t) = G_{sC}\varepsilon(t) + \eta_{effC} D_t^{\alpha}\, \varepsilon(t) + \eta_{effL} D_t^{2\alpha}\, \varepsilon(t) \text{ for regime 2}$$

where $\sigma(t)$ is the membrane stress, $\varepsilon(t)$ is the membrane strain caused by thermal fluctuations. The bulk, shear modulus of the cortex accounts for reversible (elastic) structural changes and could be expressed as:

$$G_{sC} = \frac{k_B T}{\langle l_d^2 \rangle h_C} \qquad (14)$$

while k_B is Boltzmann constant, T is temperature, $\langle l_d^2 \rangle$ is the average domain surface, h_C is the cortex thickness. The fluidity of the membrane is characterized by effective modulus: (1) η_{effC} for the cortex and (2) η_{effL} for the bilayer. The cortex rearrangement modulus, which quantifies the cortex ability to form the microdomains in the long-time regime, is related to the bulk shear modulus and expressed as:

$$B = \frac{G_{sC}}{\tau^{\alpha+1}}, \qquad (15)$$

where τ is the rearrangement time of the spectrin microdomains expressed as:

$$\tau = \left[f \left(exp\left(-\frac{\Delta G_{rear} - E\prime}{k_B T}\right) - exp\left(-\frac{\Delta G + E\prime}{k_B T}\right) \right) \right]^{-1} \quad (16)$$

where f is the characteristic frequency, ΔG_{rear} is the Gibbs energy barrier for the spectrin rearrangement, and E' is the energy perturbations caused by the membrane fluctuations (Pajic-Lijakovic and Milivojevic, 2014). The operators: $D_t^{-(\alpha+1)}(\cdot)$, $D_t^\alpha(\cdot)$, and $D_t^{2\alpha}(\cdot)$ are fractional derivatives. The fractional derivatives of a function $f(t)$ are equal to $D_t^\alpha(f(t)) = \frac{d^\alpha}{dt^\alpha} f(t)$. We used Caputo's definition of the fractional derivative (Podlubny, 1999) as follows: $D_t^\alpha(f(t)) = \frac{1}{\Gamma(1-\alpha)} \frac{d}{dt} \int_0^t \frac{f(t')}{(t-t')^\alpha} dt'$ (where t is independent variable -time and $\Gamma(1-\alpha)$ is the gamma function. If the order of the fractional derivative is $\alpha = 0$ than $D_t^0(f(t)) = f(t)$. When $\alpha = 1$, the gamma function is $\Gamma(1-\alpha) \to \infty$. For this case, the fractional derivative is not defined. However, it can be shown that when $\alpha \to 1$, then $D_t^\alpha(f(t)) \to \frac{df(t)}{dt}$. For a fractional derivative higher than zero and lower than one, this equation describes damped dissipative phenomena caused by interactions between the membrane constituents. Fractional derivatives account for the non-linear nature of the membrane viscoelasticity.

First and second terms on the right hand side of eq. 13 for both regimes represent the cortex stress while the third term is the bilayer stress. The cortex is more elastic than the bilayer and can be described by parallel coupling of two rheological elements: (1) rearrangement element and spring pot for regime 1 and (2) spring and spring pot for regime 2. The viscoelasticity of the bilayer is primarily dissipative and can be described by spring pot for both regimes. The membrane constitutive model (eq. 13) represents the product of the bilayer-cortex coupling. This coupling is expressed rheologically by Pajic-Lijakovic and Milivojevic (2014) at the macroscopic level as:

$$\sigma_L(t) \sim D_t^\alpha(\sigma_{irrC}(t)) \tag{17}$$

where $\sigma_L(t)$ is the bilayer stress (from eq. 13) equal to $\sigma_L(t) = \eta_{effL} D_t^{2\alpha} \varepsilon(t)$ and $\sigma_{irrC}(t)$ is the irreversible part of the cortex stress equal to $\sigma_{irrC}(t) = \eta_{effC} D_t^\alpha \varepsilon(t)$. The cortex-bilayer coupling under: (a) isotonic condition and (b) hypotonic conditions was shown in Figure 5.

Model eq. 13 is transformed by Fourier transformations in order to determine corresponding storage and loss moduli for both regimes. The storage modulus $G'(\omega)$ quantifies the elastic behaviour and the loss modulus $G''(\omega)$ quantifies the viscous behaviour of single erythrocytes (where ω is angular velocity). The transformed equation is $F[\sigma(t)] = G^* F[\varepsilon(t)]$, where $F[\circ]$ is the Fourier operator and G^* is the complex dynamic modulus equal to $G^* = G' + iG''$. Storage modulus is equal to:

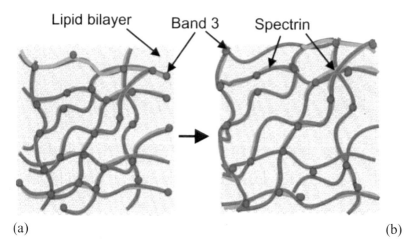

(a) (b)

Figure 5. Erythrocyte swelling under hypotonic conditions leads to the increase of the accumulated membrane stress.

$$G'(\omega) = -B \frac{1}{\omega^{\alpha+1}} \sin\left(\frac{\pi\alpha}{2}\right) + \eta_{effC} \omega^\alpha \cos\left(\frac{\pi\alpha}{2}\right) - \eta_{effL} \omega^{2\alpha}\left(1 - 2\cos^2\left(\frac{\pi\alpha}{2}\right)\right)$$
for regime 1 (18)

$$G'(\omega) = G_{sc} + \eta_{effC} \omega^\alpha - \eta_{effL} \omega^{2\alpha}\left(1 - 2\cos^2\left(\frac{\pi\alpha}{2}\right)\right)$$
for regime 2

while loss modulus is equal to:

$$G''(\omega) = -B\frac{1}{\omega^{\alpha+1}}\cos\left(\frac{\pi\alpha}{2}\right) + \eta_{effC}\omega^\alpha \sin\left(\frac{\pi\alpha}{2}\right) +$$
$$\eta_{effL}\omega^{2\alpha} 2\sin\left(\frac{\pi\alpha}{2}\right)\cos\left(\frac{\pi\alpha}{2}\right) \text{ for regime 1} \qquad (19)$$
$$G''(\omega) = \eta_{effC}\omega^\alpha \sin\left(\frac{\pi\alpha}{2}\right) + \eta_{effL}\omega^{2\alpha} 2\sin\left(\frac{\pi\alpha}{2}\right)\cos\left(\frac{\pi\alpha}{2}\right) \text{ for regime 2}$$

The model parameters: τ, η_{effC}, η_{effL}, G_{SC}, and α were obtained by comparative analyses of the four experimental data sets $G'(\omega)$ vs. ω and $G''(\omega)$ vs. ω for regimes 1 and 2 (Figure 2). The values of the parameters depend on the tonicity of external solution.

Erythrocyte swelling under hypotonic conditions induces the membrane structural changes. These structural changes are more intensive and induce higher energy dissipation for higher swelling rate obtained under lower tonicity conditions (Pajic-Lijakovic, 2015a; 2015b). The energy dissipation is quantified by increasing of the damping coefficient from α^I for isotonic to α^H for hypotonic conditions as well as by increase of the effective modulus for: (1) the cortex from η_{effC}^I to η_{effC}^H and (2) the bilayer from η_{effL}^I to η_{effL}^H (where η_{effC}^I and η_{effL}^I are the effective moduli for the cortex and the bilayer under isotonic condition, while η_{effC}^H and η_{effL}^H are the effective moduli for the cortex and the bilayer under hypotonic conditions). The thickness of the cortex decreases from h_c^I for isotonic to h_c^H for hypotonic conditions (Asami and Yamaguchi, 1999). Increase of the bilayer effective modulus during erythrocyte swelling is primarily induced by electrostatic interactions between lipids (Ashrafuzzaman and Tuszynski, 2012).

The average size of the microdomains decreases from $\langle l_d^I \rangle$ for isotonic to $\langle l_d^H \rangle$ for hypotonic conditions while the cortex rearrangement time decreases from τ^I for isotonic to τ^H for hypotonic conditions (Pajic-Lijakovic, 2015a). After erythrocyte swelling under hypotonic conditions, the membrane becomes stiffer. Consequently, the bulk shear modulus increases from G_{SC}^I for isotonic to G_{SC}^H for hypotonic conditions.

7. RESULTS AND DISCUSSION

Local changes of the bilayer bending state enhance anomalous subdiffusion and eventually lead to hop-diffusion of lipids. Cholesterol molecules stabilize changes in the lipid bilayer bending state and accumulate in high curvature area of the bilayer. It is in accordance with the fact that the cholesterol rearrangement reduces undesirable local stiffening of the bilayer which could lead to its disruption. Cholesterol spatial rearrangement could be described by characteristic time t_{chol}. This time corresponds to the lipid phase separation and the formation of stable microdomains caused by bending of the bilayer under isotonic and hypotonic conditions. Elson et al. (2010) reported that formation of stable microdomains corresponds to the time scale of seconds. The lipid phase separation is pronounced under hypotonic conditions which significantly perturbs the bending state of the bilayer during erythrocyte swelling (Leonard et al., 2017). However, band 3 molecules reduce the mobility of cholesterol (Golan et al., 1984). Stability of the bilayer after the erythrocyte swelling depends on the inter-relation between (1) the cell swelling time t_{sw} and (2) the cholesterol rearrangement time t_{chol}. Cell swelling rate has been expressed by Pajic-Lijakovic (2015a) as:

$$r_{sw} = \frac{\Delta V_e}{t_{sw}} \qquad (20)$$

where r_{sw} is the erythrocyte swelling rate and ΔV_e is the erythrocyte volume increase. The average volume of the intact human erythrocyte is $V_{e0} = 90 \pm 2\ \mu m^3$ (Nash and Meiselman, 1983). After swelling, the average volume of erythrocyte increases up to $V_{eSW} = 160\ \mu m^3$. The swelling time is $t_{sw} = 10 - 100\ s$ (Pribush et al., 2002). For faster swelling condition such that $t_{sw} < t_{chol}$, the bilayer could be disrupted. It corresponds to $t_{sw} \leq 10\ s$ (Pribush et al., 2002). For slower swelling conditions, i.e., $t_{sw} > t_{chol}$, the bilayer bending induces: (1) the disintegration of two types band 3 complexes with adducin and with spectrin (Figure 4) and (2) the increase of the freely diffusive fraction of band 3 (Golan and Veatch, 1980). The lipid-protein positive hydrophobic mismatch interactions (De Meyer et al., 2010; Standberg et al., 2012; Milovanovic et al., 2015) lead to intensive clustering of band 3 molecules and their tilting which is the prerequisite of the hemolytic hole formation. Band 3 tilting depends on the decrease of the bilayer thickness after erythrocyte swelling (Lundbaek et al., 2010). The bilayer thickness decreases up to $d_0 = d_0^H$ under hypotonic conditions, depending on the swelling rate. Tilt angle changes influence packing state of band 3 clusters and their association-dissociation to spectrin (Pajic-Lijakovic and Milivojevic, 2017). Process of the rearrangement of cholesterol is faster than the rearrangement of band 3 molecules. It is in accordance with the fact that the diffusion coefficient of the lipids is approximately 4 times greater than the diffusion of the band 3 molecules in the same membrane (Golan and Veatch, 1980). Consequently, the lifetime of the lipid structural integrity corresponds to the time for band 3 rearrangements before the reversible hemolytic hole formation (Pribush et al., 2002; Pajic-Lijakovic, 2015a). The band 3 lateral motion depends on the cortex structural ordering (Tomishige et al., 1998). This ordering is induced by coupling with the lipid bilayer (Pajic-Lijakovic and Milivojevic, 2014). An excited cluster of band 3 molecules (caused by the bilayer-cortex coupling) changes its packing state from dense packing to the ring-like structure and forms the hemolytic hole (Sato et al., 1993; Pajic-Lijakovic et al., 2010). Seeman et al. (1973) experimentally determined the diameter of the reversible hemolytic holes in the range between 10 nm to 100 nm for human erythrocyte under hypotonic conditions at pH=7. Consequently, the hole can be formed even from small clusters which consist of 4 molecules of band 3 (Pajic-Lijakovic et al., 2010; Pajic-Lijakovic, 2015a). Spatial distribution of band 3 molecules and their clustering are influenced by bending state of the lipid bilayer.

The bilayer coupling with the actin-spectrin cortex influences the conformational changes of spectrin filaments (Gov and Safran, 2005; Li et al., 2005). Cumulative effects of these changes influence the anomalous nature of the erythrocyte membrane viscoelasticity (Pajic-Lijakovic and Milivojevic, 2014). The spectrin filament conformations are dependent on the state of three types of the complexes (Figure 4): (1) band 3 complexes with ankyrin, (2) band 3 complexes with adducin, and (3) band 3 complexes with spectrin (Pajic-Lijakovic, 2015b). The flexibility of the spectrin filaments depends on the number of band 3 molecules attached per single filaments.

The fastest sub-process which influneces the membrane adaptation under hypotonic conditions is the Hb release through the reversible hemolytic hole. This release occurs during successive open-closed cycles of the hole. Zade-Open (1998) reported that the average open time period corresponds to 270 milliseconds while the average closed time period corresponds to 260 milliseconds. The release time period is significantly shorter than the erythrocyte swelling time under hypotonic conditions (Pribush et al., 2002). This condition represents a necessary condition in order to keep the membrane integrity.

The hole opening is connected with the perturbation of the state of band 3 cluster from dense packing to ring-like structure (Pajic-Lijakovic et al., 2010). This change of the cluster state also perturbes the state of surrounding lipids.

Consequently, the cell swelling rate is the key factor which determines the whole process under hypotonic conditions (Pribush et al., 2002; Pajic-Lijakovic, 2015a). Fast swelling rate significantly reduces the lifetime of the membrane integrity as well as whole hemolysis time. However, a decrease in the swelling rate induces an increase of the membrane lifetime as well as the hemolysis time. If the swelling rate is slow enough, the membrane has sufficient time to relax. Under these conditions, the hemolytic hole doesn't have to appear (Pribush et al., 2002). Consequently, the cell swelling rate could serve as the process parameter for the optimization of the osmotic hemolysis.

CONCLUSION

The lipid bilayer bending induces its stiffening. Cholesterol accumulation within the bending bilayer part could reduce this stiffening which is the prerequisite for ensuring the membrane integrity. The bilayer bending influences: (1) the spatial distribution of band 3 molecules and (2) the rearrangement of spectrin filaments directly - through the bilayer-cortex coupling and indirectly - by changing the number of band 3 molecules attached to spectrin. The flexibility of the spectrin filaments can vary from flexible to even rigid depending on the number of attached band 3 molecules per single spectrin filament. Cholesterol rearrangement influences band 3 clustering and also can induce the protein tilting. However, the cholesterol lateral diffusion can be reduced by interactions with band 3 molecules through positive hydrophobic mismatch effects.

These inter-relations between the main membrane components depend on the rate of the bilayer bending. Thermal fluctuations of the erythrocyte membrane induce local bending of the bilayer. However, cell swelling under hypotonic conditions significantly changes the bilayer bending state. The rate of erythrocyte swelling influences the lifetime of bilayer integrity. The faster swelling condition cannot ensure enough time for the cholesterol rearrangement which can lead to the undesirable bilayer disruption. However, slower swelling prevents the bilayer stiffening by the cholesterol rearrangement and ensures a prolonged lifetime of the bilayer integrity. These changes of the bilayer state influence the rearrangement of band 3 molecules and their clustering which can result in reversible hemolytic hole formation. The whole process of osmotic hemolysis can be regulated by changing the erythrocyte swelling rate depending on the ionic strength of the external solution.

ACKNOWLEDGMENTS

The authors gratefully acknowledge funding support of the Ministry of Education, Science and Technological Development of the Republic Serbia (grants III 46001).

REFERENCES

Almeida, P.F.F. (2009) Thermodynamics of lipid interactions in complex bilayers. *Biochimica et Biophysica Acta* 1788:72-85.

Amin, M.S., Park, Y.K., Lue, N., Dasari, R.R., Badizadegan, K., Feld, M.S., Popescu, G. (2007) Microrheology of red blood cell membrane using dynamics scattering microscopy. *Optics Express,* 15(25):17001-17009.

Argudo, D., Bethel, N.P., Marcoline, F.V., Grabe, M. (2016) Continuum descriptions of membranes and their interaction with proteins: towards chemically accurate models. *Biochimica et Biophysica Acta* 1858(7): 1553-1790.

Asami, K., Yamaguchi, T. (1999) Electrical and Morphological Changes of Human Erythrocytes under High Hydrostatic Pressure Followed by Dielectric Spectroscopy. *Annals of Biomedical Engineering* 27:427-435.

Ashrafuzzaman, Md.; Tuszynski, J. (2012) Regulation of Channel Function Due to Coupling with a Lipid Bilayer. *Journal of Computational and Theoretical Nanoscience,* 9(4):564-570.

Boal, D. (2012) *Mechanics of the Cell.* Second edition, Cambridge University Press, New York.

Blackman, S.M., Cobb, C.E., Beth, A.H., Piston, D.W. (1996) The Orientation of Eosin-5-Maleimide on Human Erythrocyte Band 3 Measured by Fluorescence Polarization Microscopy. *Biophysical Journal* 71:194-208.

Choubey, A., Kalia, R.K., Malmstadt, N., Nakano, A., Vashishta, P. (2013) Cholesterol Translocation in Phospholipid Membrane. *Biophysical Journal* 104:2429-2436.

De Meyer, F.J.M., Rodgers, J.M., Willems, T.F., Smit, B. (2010) Molecular Simulation of the Effect of Cholesterol on Lipid-Mediated Protein-Protein Interactions. *Biophysical Journal* 99:3629-3638.

Destainville N (2008) *Cluster phases of membrane proteins.* arXiv:cond-mat/0607400v2 [cond-mat.soft] 1-5.

Elson, E.L., Fried, E., Dolbow, J.E., Genin, G.M. (2010) Phase separation in biological membranes: integration of theory and experiment. *Annual Review of Biophysics* 39:207-226.

Franco, T., Low, P.S. (2010) Erythrocyte adducin: A structural regulator of the red blood cell membrane. *Transfusion and Clinical Biology* 17(3):87-94.

Fujiwara, T., Ritchie, K., Murakoshi, H., Jacobson, K., Kusumi, A. (2002) Phospholipids undergo hop diffusion in compartmentalized cell membrane. *Journal of Cell Biology* 157(6):1071-1081.

Fujiwara, T.K., Iwasawa, K., Kalay, Z., Tsunoyama, T.A., Watanabe, Y., Umemura, Y.M., Murakoshi, H., Suzuki, K.G.N., Nemoto, Y.L., Morone, N., Kusumi, A. (2016) Confined diffusion of transmembrane proteins and lipids induced by the same actin meshwork lining the plasma membrane. *Molecular Biology of the Cell.* 27:1101-1119.

Gil, T., Ipsen, J.H., Mouritsen, O.G., Sabra, M.C., Sperotto, M.M., Zuckermann, M.J. (1998) Theoretical analysis of protein organization in lipid membranes. *Biochemica et Biophysica Acta* 1376:245-266.

Golan, D.E., Alecio, M.R., Veatch, W.R., Rando, R.R. (1984) Lateral Mobility of Phospholipids and Choloesterol in the Human Erythrocyte Membrane: Effects of Protein-Lipid Interactions. *Biochemistry* 23:332-339.

Golan, D.E., Veatch, W. (1980) Lateral mobility of band 3 in the hyman erythrocyte membrane studied by fluorescence photobleaching recovery: Evidence for control by cytoskeletal interactions, *PNAS* 77(5):2537-2541.

Gov, N.S., Safran, S.A. (2005) Red Blood Cell Membrane Fluctuations and Shape Controlled by ATP-Induced Cytoskeletal defects. *Biophysical Journal.* 88:1859-1874.

Kabaso, D., Shlomovitz, R., Auth, T., Lew, V.L., Gov, N.S. (2011) Cytoskeletal Reorganization of Red Blood Cell Shape: Curling of Free Edges and Malaria Merozoites. In: (ed.) Iglic A., *Advances in Planar Lipid Bilayers and Liposomes,* Volume 13, pp. 73-102.

Kheyfets, B., Galimzyanov, T., Drozdova, T., Mukhin, S. (2016) Analytical calculation of the lipid bilayer bending modulus. *Physical Review* E 94:042415 1-11.

Kodippili, G.C., Spector, J., Hale, J., Giger, K., Hughes, M.R., McNagny, K.M., Birkenmeier, C., Peters, L., Ritchie, K., Low, P.S. (2012) Analysis of the Mobilities of Band 3 Populations Asociatted with Ankyrin Protein and Junctional Complexes in Intact Murine Erythrocytes. *Journal of Biological Chemistry* 287(6):4129-4138.

Leonard, C., Conrad, L., Guthmann, M., Pollet, H., Carquin, M., Vermylen, C., Gailly, P, Van Der Smissen, P.,Mingeot-Leclercq, M.P., Tyteca, D. (2017) Contribution of plasma membrane lipid domains to red blood cell (re) shaping. *Science Report* 7:4264 1-17.

Leiber, M.R., Steck, T.L. (1982) A description of the holes in human erythrocyte membrane ghosts. *Journal of Biological Chemistry* 257:11651-11659.

Li, J., Dao, M., Lim, C.T., Surech, S. (2005) Spectrin-level Modeling of the Cytoskeleton and Optical Tweezers Stretching of the Erythrocyte. *Biophysical Journal* 88:3707-3719.

Lundbaek, J.A., Collingwood, S.A., Ingolfsson, K.I., Kapoor, R., Andersen, O.S. (2010) Lipid bilayer regulation of membrane protein function: gramicidin channels as molecular force probes. *Journal of the Royal Society Interface* 7:373-395.

Mikhalyov, I. and Samsonov, A. (2011) Lipid raft detecting in membranes of live erythrocytes. *Biochimica et Biophysica Acta.* 1808:1930-1939.

Milovanovic, D., Honigmann, A., Koike, S., Gottfert, F., Pahler, G., Junius, M., Muller, S., Diederichsen, U., Janshoff, A., Grubmuller, H., Risselada, H.J., Eggeling, C., Hell, S.W., van den Bogaart, G., Jahn, R. (2015) Hydrophobic mismatch sorts SNARE proteins into distinct membrane domains. *Nature Communication* 6:5984 1-10.

Nash, G.B., Meiselman, H.J. (1983) Red Cell and Ghost Viscoelasticity, Effects of Hemoglobin Concentration and In Vivo Aging. *Biophysical Journal* 43:63-73.

Podlubny, I. (1999) *Fractional Differential Equations, Mathematics in Science and Engineering.* London: Academic Press, 198 pp. 78.

Pajic-Lijakovic, I., Ilic, V., Bugarski, B., Plavsic, M.B. (2010) The rearrangement of erythrocyte band 3 molecules and reversible osmotic holes formation under hypotonic conditions. *European Biophysics Journal and Biophysics Letters* 39(5):789-800.

Pajic-Lijakovic, I., Milivojevic, M. (2014) Modeling analysis of the lipid bilayer-cytoskeleton coupling in erythrocyte membrane. *Biomechanics and Modeling in Mechanobiology* 13(5): 1097-1104.

Pajic-Lijakovic, I. (2015a) Erythrocytes under osmotic stress–modeling considerations. *Progress in Biophysics and Molecular Biology* 117(1):113-124.

Pajic-Lijakovic, I. (2015b) Role of Band 3 in Erythrocyte Membrane Structural Changes under Thermal Fluctuations-Modeling Considerations. *Journal of Bioenergetics and Biomembranes* 47(6):507-518.

Pajic-Lijakovic, I., Milivojevic, M. (2017) Role of Band 3 in the Erythrocyte Membrane Structural Changes Under Isotonic and Hypotonic Conditions, (ed. Jimenez-Lopez, J.C.) In: *Cytoskeleton - Structure, Dynamics, Function and Disease,* InTech, pp. 89-103.

Popescu, G., Park, Y.K., Dasari, R.R., Badizadegan, K., Feld, M.S. (2007) Coherence properties of red blood cell membrane motion. *Phys. Rev. E* 76, 031902 1-5.

Pribush, A., Meyerstein, D., Meyerstein, N. (2002) Kinetic of erythrocyte swelling and membrane hole formation in hypotonic media. *Biochimica et Biophysica Acta* 1558:119-132.

Puig-de-Morales-Marinkovic, M., Turner, K.T., Butler, J.P., Fredberg, J.J., Suresh, S. (2007) Viscoelasticity of the human red blood cell. *American Journal of Physiology and Cell Physiology* 293: C597–C605.

Sato, Y., Yamakose, H., Suzuki, Y. (1993) Mechanism of Hypotonic Hemolysis of Human Erythrocytes. *Biol. Pharm. Bull.* 16(5):506-512.

Saxton, M.J. (1990) Lateral diffusion in a mixture of mobile and imobile particles; A Monte Carlo study. *Biophysical Journal* 58:1303-1306.

Seeman, P., Cheng, D., Iles, G.H. (1973) Structure of membrane holes in osmotic and saponian hemolysis. *Journal of Cell Biology* 56:519-527.

Sens, P., Turner, M.S. (2004) Theoretical Model for the Formation of Caveolae and Similar Membrane Invaginations, *Biophysical Journal* 86:2049-2057.

Shlomovitz, R., Gov, N.S. (2008) Curved inclusions surf membrane waves. *Europhysics Letter.* 84:58008 p1-p6.

Simons, K., Toomre, D. (2000) Lipid rafts and signal transduction. *Nature Review and Molecular and Cell Biology* 1(1):31-39.

Strandberg, E., Esteban-Martin, S., Ulrich, A.S., Salgado, J. (2012) Hydrophobic mismatch of mobile transmembrane helices: merging theory and experiments. *Biochimica et Biophysica Acta.* 1818:1242-1249.

Taylor, A.M., Boulter, J., Harding, S.E., Colfen, H., Watts, A. (1999) Hydrodynamics Properties of Hyman Erythrocyte Band 3 Solubilized in Reduced Triton X-100. *Biophysical Journal* 76:2043-2055.

Trimbe, W.S. and Grinstein, S. (2015) Barriers of the free diffusion of proteins and lipids in the plasma membranes. *Journal of Cell Biology.* 208(3):259-271.

Tomishige, M., Sako, Y., Kusumi, A. (1998) Regulation mechanism of the lateral diffusion of band3 in erythrocyte membranes by the membrane skeleton. *Journal of Cell Biology* 142(4):989-1000.

Zade-Oppen, A.M.M. (1998) Repetitive cell 'jump' during hypotonic lysis of erythrocytes observed with simple flow chamber. *Journal of Microscopy* 192:54-62.

In: Lipid Bilayers
Editor: Mohammad Ashrafuzzaman

ISBN: 978-1-53616-392-6
© 2019 Nova Science Publishers, Inc.

Chapter 4

MICROBIAL-DERIVED BIOACTIVE LIPOPEPTIDES: PORE-FORMING METABOLITES IN LIPID BILAYERS

Yulissa Ochoa, Mariana Bernal and Daniel Balleza[*]
Department of Chemistry ICET,
Universidad Autónoma de Guadalajara,
Zapopan, Jalisco, México

ABSTRACT

Microbial cyclic lipopeptides are considered natural biosurfactants due to their amphipathic structures. They exhibit interesting antibiotic properties mainly related to their interactions with lipid membranes. Understanding how mechanical properties of lipid bilayers contribute and determine the antagonistic activity of these secondary metabolites over a broad spectrum of microbial pathogens could establish a framework to propose more efficient strategies of biological control. This implies unravelling at the biophysical level the complex interactions established between these amphiphilic molecules and lipid bilayers. Our review aims

---

[*] Corresponding Author's Email: daniel.balleza@edu.uag.mx.

to update this knowledge and discuss its relevance for the agronomic and pharmaceutical industries.

LIPID ABBREVIATIONS

bSM	brain sphingomyelin
DiPhyPC	1,2-diphytanoyl-*sn*-glycero-3-phosphocholine
DMPC	1,2-dimyristoyl-*sn*-glycero-3-phosphocholine
DPPC	1,2-dipalmitoyl-*sn*-glycero-3-phosphocholine
DPPE	1,2-Dipalmitoyl-*sn*-glycero-3-phosphoethanolamine
DPPS	1,2-Dipalmitoyl-*sn*-glycero-3-phosphoserine
DSPC	1,2-distearoyl-*sn*-glycero-3-phosphocholine
POPC	1-palmitoyl-2-oleoyl-*sn*-glycero-3-phosphocholine
POPG	1-palmitoyl-2-oleoyl-*sn*-glycero-3-phosphoglycerol

1. INTRODUCTION

Lipopeptides (LPs) are amphipathic compounds with biosurfactant properties that are synthesized as secondary metabolites by several microorganisms. Notably, in addition to their surfactant properties, LPs also exhibit pharmacological activities including antimicrobial, antitumor and immunosuppressant properties (Koglin et al. 2010; Biniarz et al. 2017). They contain a hydrophilic region usually composed of amino acids and a hydrophobic region of fatty acyl nature. LPs from Bacillaceae genera (for example *Bacillus* and *Paenibacillus*) are divided into three families (surfactin, fengycin, and iturin) according to the structure of the cyclic peptides and they are the most widely studied. Given their amphipathic structures and bioactivities, these compounds are of high interest for the development of tools to diverse industries, including the agricultural and pharmaceutical ones. Indeed, the anti-fungal activity of some LPs is the most widely used property of these compounds, not only in feed and plant defence but also in food preservation. Other bio-activities of LPs are being developed for pharmaceutical applications, such as their use as antibiotics (Cochrane and Vederas 2016). These and other biotechnological applications are especially promising due to the wide anti-microbial spectrum of these compounds, making them more difficult to produce resistance. However, to fully understand how LPs act, it is crucial to unravel how they interact with lipid bilayers. Due to the complexity of biological systems, advances in the biophysics of their mechanisms of action have contributed to exploit their potentialities. Model membranes (monolayers, supported bilayers, and liposomes) have been broadly used to characterise several bioactive compounds in relation to their mechanical and thermodynamical properties, lipid composition and the lipid/lipid interfaces. Although these models do not reflect the complexity of biological membrane structure, their study has demonstrated some important facts to understand the bio-activities of those molecules. Thus, the combination of different lipid biophysical techniques using several membrane models provides a comprehensive and detailed framework to analyse the mechanisms of membrane perturbation by a plethora of linear and cyclic LPs. Here, the main unravel biophysical mechanisms of action of several cyclic LPs are reviewed and some strategies to use these compounds as promising antibiotics in the pharmaceutical industry and particularly in agrobiotechnology are discussed.

1.1. Ecological Significance of Endophytic Associations between Plants and Microbia

Plants are naturally associated with several kinds of microorganisms in diverse ways. Endosymbiotic beneficial interactions, such as the root nodule symbiosis of legumes with rhizobia or the formation of arbuscular mycorrhiza with fungi are accommodated intracellularly and surrounded by a host membrane. On the other hand, in pathogenic interactions, bacteria often inject effector proteins into plant host cells to overcome defense responses, mechanisms which are also utilized by symbionts. Endophytic bacteria and fungi are microorganisms living inside plant tissues without causing damage symptoms to the plant and frequently these associations are a source of beneficial bioactive compounds (Brader et al. 2014; Wani et al. 2015; Singh et al. 2017) (Figure 1). In contrast to what is frequently observed in endosymbionts, endophytes are not residing in living plant cells or surrounded by internal membranes but the molecular basis of these interactions is not well understood. Thus, the plant root is a complex surface (biofilm) with an exotic chemistry, capable of directing various interactions with the microbial community. The chemical complexity of root secretions implies the existence of complex mechanisms facilitating beneficial and suppressing pathogenic biofilms. As biosurfactants, LPs are involved in biofilms formation, root colonization and they are also crucial for biocontrol activity and systemic resistance in plants (Kolter et al. 2010).

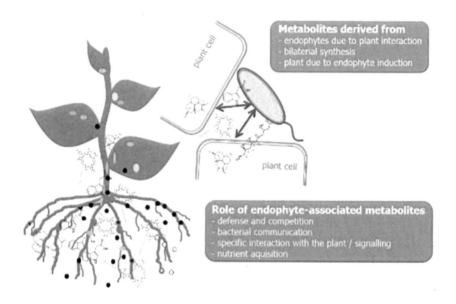

Figure 1. Cross-talk between plants and endophytic bacteria. Some endophytes are known to produce several antibiotic compounds such as cyclic lipopeptides to protect host plants from diverse pathogens (Courtesy of Dr. A. Sessitsch, adapted from Brader et al. 2014).

1.2. The Lipid Bilayer: Action Target of Lipopeptide Compounds

Biological membranes are composed by a lipid bilayer to which several proteins, carbohydrates and modified lipids are associated by non-covalent or covalent links. The lipid bilayer is also a molecular matrix to establish various interactions of a purely mechanical nature. Every cell responds, to a greater or lesser degree, to the mechanical stimuli of its environment. This class of phenomenology can be considered, therefore, as one of the most ancestral, probably universal, in every organism. Evolutionarily, the mechanosensitive ion channels (MS) have developed the specific ability to respond, in a regulated manner, to this kind of stimuli (Kung 2005). However, diverse peptides, proteins and membrane bioactive compounds also respond to this class of stimuli (Teng et al. 2015). Hence, in order to better understand the function of any interacting peptide or protein on lipid membranes, it is necessary to study the physical properties that, as a material, have the lipid matrix (Phillips et al. 2009).

The lipid component of biological membranes determines, for example, the distribution, as well as the organization and proper function of several membrane proteins, importantly, MS channels. Moreover, they determine the interaction with external molecules such as antimicrobial peptides (AMPs), LPs, and several types of ligands. In general terms the activity of these biomolecules is the result of specific lipid-peptide interactions, general bilayer-protein interactions and, interactions in the hydrophobic-aqueous milieu interface. Such interactions occur mainly because the coupling between the hydrophobic phase of the bilayer with respect to the hydrophobic domains of these biomolecules and, in consequence, are a function of the material properties of the bilayer as a whole, including factors such as its thickness, degree of compaction, intrinsic curvature, elasticity, rigidity, rheological properties, asymmetry, surface charge and chemical composition (McIntosh and Simon 2006; Andersen and Koeppe 2007; Ashrafuzzaman et al. 2014). Therefore, lipid lateral segregation depends on these factors and it is modulated by temperature and lateral tension (see below). This determines the segregation and formation of lipid domains, which are of great relevance in biological terms (Figure 2). In this sense, the study of the form and dynamics of such domains, determined by the lateral surface tension and the interaction between the lipid dipoles, is important for understanding lipid-protein interactions and its relevance in protein function. It is well known that some lipids are segregated laterally in 'rafts' of solid nature (S_o) or liquid ordered (L_o), while other ones generally prefer a disordered liquid phase (L_d or $L_α$) of high fluidity, which facilitates the conformational changes that certain membrane proteins carry out during the performance of its activity (Goñi 2014).

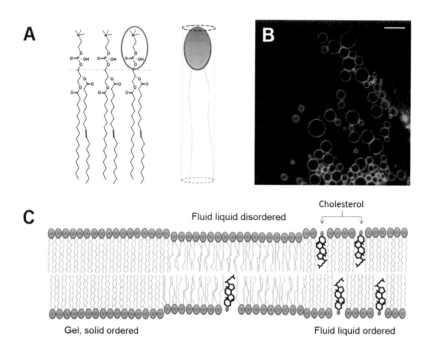

Figure 2. Lipids and lipid bilayers. (A) Chemical structure of POPC (1-palmitoyl-2-oleoyl-sn-glycero-3-phosphocholine). Lipids are characterized by their *critical packing parameter* (v/la_0) where v = nonpolar volume; l = nonpolar length and a_0 = optimal surface area of the polar headgroup. (B) Self-assembly of lipids in contact with water forming unilamellar vesicles (bar ≈ 40 μm). (C) Lipid bilayer phases. Depending on lipid composition, temperature and packing parameters, the lamellar phase can adopt solid or fluid phases. As temperature decreases, the bilayer passes from a liquid-disordered phase (L_d) through a gel phase (S_o). The L_d phase is highly dynamic. In presence of cholesterol, lipid bilayers can adopt an extra lamellar phase, the liquid-ordered (L_o) phase, which shares the characteristics of both gel and fluid ones.

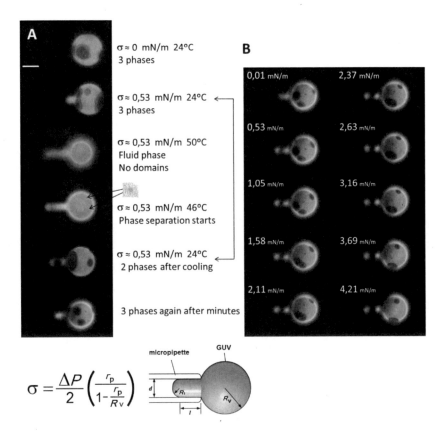

Figure 3. Three-Phase lipid coexistence depends on temperature and lateral tension (σ). (A) Effect of the temperature at constant lateral tension for a GUV of DiPhyPC/bSM/dihydrocholesterol (molar ratio 1:1:1). The gray region disappears at high temperature (>49°C) and it is reformed at low temperature (<24°C) once the system acquires equilibrium. (B) Effect of the lateral tension at constant temperature and evolution of the intermediate phase upon increasing lateral tension (σ). The value of the tension is reported in each image for the same vesicle showing coalescence of the intermediate domain by partially overlapping circular DPhyPC-rich (L_d) and bSM-rich (L_o) domains. *Note*: No significant differences were found using cholesterol. Bar ≈ 20 μm.

In Figure 3 a lipid model system, mimicking the typical composition of a mammalian cell, constituted by a giant unilamellar vesicle (GUV) composed of a ternary mixture of lipids (DiPhyPC/bSM/dihydrocholesterol 1:1:1 molar ratio) exhibits three fluorescence intensity levels corresponding to three different physical states of the lipid membrane, namely L_o, L_d and an intermediate phase. This liposome was microaspirated at room temperature and it is noteworthy that by increasing the lateral tension, the intermediate region is completely transformed to a region characterized by a darkest intensity, implying a deep reorganization of the lipid molecular matrix. The relevance of this is that organization of lipid membranes is strongly dependent on mechanical properties and purely physical variables such as temperature and lateral tension (Balleza et al. 2019).

The microaspiration technique (MAT) is a simple method to estimate liposome stiffness by measuring the degree of membrane deformation in response to negative pressure applied by a glass micropipette to the lipid bilayer surface. These experiments not only allow making measurements of liposomes mechanical properties, but they also allow insight into the mechanisms of interaction between lipid bilayers and peptides, ligands and several pharmaceuticals. The response to the aspiration pressure is the liposome's deformation through a hemi-spherical projection inside the pipette. Depending on lipid composition, beyond certain point, the applied pressure difference can be converted to lateral tension (σ or τ) when a liposome reaches a new equilibrium position. Thus, surface tension and the elastic stretching modulus (Ka or K_S) can be calculated according the Laplace law. In a typical microaspiration experiment, the lateral tension (τ) is a function of the pressure difference (ΔP) in the bilayer (in N/m), r_p is the internal diameter of the micropipette and R_V is the external radius of the vesicle (see *scheme* in Figure 3).

Microaspiration and atomic force microscopy (AFM) have been used to study the mechanism of action of several AMPs and LPs (Etchegaray et al. 2008; Lee et al. 2008, 2018). In addition to microaspiration and AFM, there have other biophysical methods to get information on the molecular effects of LPs and AMPs on lipid bilayers. These include (among others techniques): the Planar Lipid Bilayer system (PLB), the Langmuir monolayer technique, differential scanning calorimetry (DSC), isothermal titration calorimetry (ITC), fluorescence spectroscopy and imaging, neutron reflectivity (NR), electron paramagnetic spectroscopy (EPR), nuclear magnetic resonance (NMR), electron paramagnetic resonance (EPR), oriented linear and circular dichroism (CD) as well as infrared spectroscopy

(FTIR). As LPs, AMPs are interesting molecules which also possess antibacterial and antifungal activities and in addition are probably the best pore-inducing molecules in nature (Marin-Medina et al. 2016, Huang and Charron 2017). Thus, due to the evident similarity between them, mechanisms of action by which these compounds exert their effects are often analogous and generally include two types of alterations of the lipid bilayer: (*i*) destabilization of the lipid matrix and (*ii*) pore formation or detergent-like solubilisation. In the following sections, the main evidences that have been described in the literature regarding the molecular effects exerted by LPs on lipid bilayers will be discussed. The reader interested in the effects described for AMPs will find excellent reviews already published with a large amount of information on the subject (Fuertes et al. 2011; Marquette and Bechinger 2018; Moravej et al. 2018).

2. Cyclic Lipopeptides from *Bacillus*: Surfactins, Iturins and Fengycins

The study of microbial biofilms has provided important clues about how they have protecting and enhancing properties by agriculturally beneficial microorganisms. In this context, a wide range of structurally different biosurfactants have been identified, including lipopeptides, glycolipids, polypeptides, polysaccharides, and lipoproteins, or mixtures thereof (Neu, 1996; Ron and Rosenberg, 2001; Mulligan, 2005; Abdel-Mawgoud and Stephanopoulos 2017). LPs are comprised of a hydrophobic tail, which is usually a fatty acid, linked to a hydrophilic linear or cyclic head of 4 – 12 amino acids via ester or amide bonds or both. Often the amino acids are of the D- rather than the usual L-configuration, presumably to resist the action of proteases (Hamley 2015). They are produced by different microorganisms, such as bacteria, yeasts, and filamentous fungi, and are generally resistant to temperature and enzymatic hydrolysis. These characteristics make them excellent biotechnological options for the development of novel antibiotics (Fiechter 1992). The best known cyclic

LPs are produced by several bacteria and fungi, including genera such as *Bacillus, Pseudomonas, Streptomyces*, and *Aspergillus*. They have several properties including antimicrobial activity against pathogenic bacteria, fungi and viruses, involvement in bacterial motility, facilitation of swarming behaviour as well as attachment to surfaces. The best studied LPs include surfactin, iturins, bacillomycins, fengycins, kurstakin, daptomycin, viscosins, syringomycins and syringopeptins (Fiechter 1992; Malev et al. 2002; Hamley 2015; Mnif 2015; Taylor and Palmer 2016; Biniarz et al. 2017).

2.1. Surfactin and Iturins

Bacillus is a gram-positive bacterial genus, member of the phylum Firmicutes which has been widely recognized as a producer of different LPs, among them surfactin (SF), iturins (It) and fengycins (FG) (Meena and Kanwar 2015; Perez 2017) (Figure 4). SF (~1.36 kDa) is an amphipathic cyclic lipoheptapeptide with the peptide ring interlinked with β-hydroxyfatty acids with chain lengths of 12 to 16 carbons to form a cyclic lactone ring structure (Figure 5). The interaction of SF with lipid membranes involves individual insertion followed by pore formation and micellation which contributes to solubilize lipid membranes (Heerklotz and Seelig 2001). This behaviour is dependent on the membrane dipole potential (Ostroumova et al. 2010) and is strongly dependent on lipid composition (Carrillo et al. 2003; Heerklotz 2007). Indeed, SF preferably interacts with negatively charged lipids (Buchoux et al. 2008), while cholesterol is reported as an attenuator of the membrane-perturbing effect of SF (Carrillo et al. 2003). Importantly, SF and other LPs achieve their molecular actions in a concentration-dependent manner and it is lipid phase-sensitive (Deleu 2013) or it induces exotic lipid phases such as the evasive ripple phase at temperatures below the pretransition of dipalmitoyl-phosphatidylcholine (DPPC) bilayers (Brasseur et al. 2007).

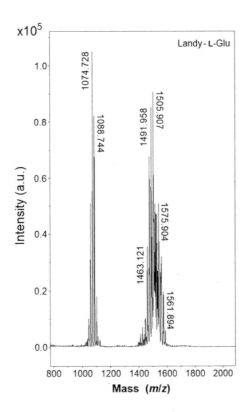

Figure 4. Matrix-assisted laser desorption ionization–time of flight (MALDI-TOF) mass spectrum of cell-surface extracts obtained from *Bacillus tequilensis* grown in modified Landy's medium (Landy et al. 1948). Peaks of putative homologs for surfactin (m/z 1000–1100); iturin homologs (m/z 1066–1110); bacillomycin (m/z 1030–1111) and [M+H]$^+$ as well as [M+Na]$^+$ or [M+K]$^+$ adducts of fengycin and plipastatin (m/z 1450–1544) indicate that this strain produce complex mixtures of all these lipopeptides. Under other growth conditions, we also detected a set of m/z ratios that we putatively ascribed to isomers of kurstakin (m/z ratios from 906.50 to 944.50, El Arbi et al. 2016) (*data not shown*). Mass spectra were obtained according previous parameters (Beltran-Gracia et al. 2017).

Actions of SF also depend on acyl-chain length as well as headgroup structure with high miscibility for shorter chain lengths (*i.e.*, DMPC>DPPC>DSPC) and polar headgroup in the follow order: DPPC>DPPE>DPPS (Bouffioux et al. 2007). Regarding the peptidic ring, which is a cyclic heptapeptide moiety with the sequence Glu-Leu-D-Leu-Val-Asp-D-Leu-Leu linked to a C14-15 β-hydroxy fatty acid (Heerklotz and

Seelig 2001, 2007), it has been demonstrated that membrane activity of SF strongly depends on geometry and composition of this structure, its charge, as well as the aliphatic chain length (Deleu et al. 1999; Eeman 2005, 2006, 2009a, 2009b; Francius et al. 2008). In sum, the actions of SF can be summarized by a destabilizing and permeabilizing effect on membranes via a so-called detergent-like action.

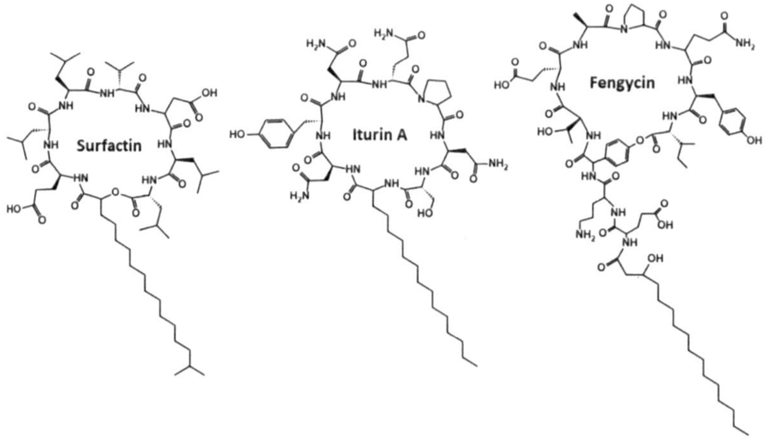

Figure 5. Microbial cyclic lipopeptides. The structures of representative members of surfactin (amino acid ring: L-Glu – L-Val – D-Leu – L-Ala – L-Asp – D-Leu – L-Val; acyl chain C12 – C16), iturin-A (amino acid ring: L-Asn – D-Tyr – D-Asn – L-Gln – L-Pro – D-Asn – L-Ser; acyl chain C14 – C17) and fengycin (amino acid ring: L-Glu – D-Orn – D-allo-Thr – L-Glu – D-Ala – L-Pro – L-Gln – L-Tyr – L-Ile; acyl chain C14 – C18) synthesized by Bacillus species. These cyclic lipopeptides contain fatty acid chain linked with amino acids forming a ring (see text). The derivatives of compounds in each group come from different amino acid components. In fengycins, double bond between carbons 2-3, 3-4 or 13-14 in acyl chains have been reported. Orn = ornithine.

Among the three main types of LPs, iturin is comparatively the smaller one with a molecular mass of ~1.1 kDa. Iturin A (It-A) is a cyclic heptapeptide linked to a fatty acid (β-amino) chain that can vary from C-14 to C-17 carbons (Figure 5). It-A forms conducting pores with anion selectivity in planar bilayers. The channels formed by these LPs strongly depend on the lipid composition, particularly on cholesterol (Maget-Dana 1985; Harnois 1989). Synergistic action has been reported in It-A/SF mixtures (Razafindralambo et al. 1997). Their mechanism of action has not been completely understood but there is experimental evidence indicating that iturins form aggregates with phospholipids and vesicularization. This destabilizes lipid membranes (Thimon et al. 1995; Grau et al. 2000). Iturins also form aggregates with SFs when both are present. On the other hand, *in vivo* assays have shown that iturins disturb the plasma membrane promoting leakage of many cellular components, mitochondrial membrane disorganization, and inhibition of the respiratory rate through a mechanism still poorly understood (Gordillo et al. 2015).

2.2. Fengycin

Fengycin (FG) (~1.46 kDa) is a lipodecapeptide containing lactone ring in the β-hydroxy fatty acid chain from C-14 to C-18 carbons that can be saturated or not (Figure 5), giving different homologous compounds and isomers. On the basis of variation at single amino acid residue at the 6th position in the ring, FGs have been classified in two classes, namely, FG-A and FG-B. FG-A contains Ala at this position but it is replaced by Val in FG-B (Vanittanakom et al. 1986). FG is ineffective against yeast and bacteria but it inhibits filamentous fungi. The *mechanism* of *FG action* is related with its surface activity which strongly depends on lipid composition and several environmental conditions, for example, temperature, pH, ionic strength and hydration (Eeman 2005; 2009b). As iturins, FG induces important membrane perturbations, including changes in the compressibility properties of DPPC monolayers by locally forming large aggregates. In other words, FG interferes with the tight packing of phospholipid molecules which is required for gel-phase formation, having a similar fluidizing effect as for cholesterol (Deleu et al. 2005). The actions of FG also depends on the concentration and it is based on a two-state transition: a monomeric state, not deeply anchored, and a buried, aggregated state, which is responsible for membrane leakage through micellization of lipid membranes, increasing bilayer fluidity and promoting membrane positive curvature (Deleu et al. 2008; Horn et al. 2013). Interestingly, FG is able to dehydrate the polar surface in DPPC membranes indicating loss in hydrogen-bonding at the headgroups, which rigidifies the lipid/water interface and decreases both the motional freedom of the phospholipid. Thus, due to iits amphipathic character, FG modifies the barrier properties of lipid bilayers (González-Jaramillo 2017).

2.3. Daptomycin

Daptomycin (DAP) (~1.62 kDa) is a 13 amino acid, cyclic lipopeptide produced by a non-ribosomal peptide synthetase mechanism in *Streptomyces roseosporus*. Ten of the 13 amino acids are arranged in a ring, and three residues are exocyclic linked to a 10-carbon lipophilic tail (Figure 6). Unlike the LPs produced by endophytic *Bacillus*, its main application is in the clinical and pharmaceutical industry since this compound has important antibiotic activity against Gram-positive pathogens, including vancomycin- and methicillin resistant *Staphylococcus aureus*, penicillin-resistant *Streptococcus pneumoniae*, *S. agalactiae*, *S. dysgalactiae* ssp. *equisimilis*, and *S. pyogenes*, vancomycin-resistant *Enterococcus faecium* and other antibiotic resistant strains (Steenbergen et al. 2005; Kanafani and Corey 2007). The atypical structure of DAP determines a novel mechanism of action where multiple aspects of the bacterial cell membrane function are disrupted. Due to its amphipathic character, DAP insert its lipophilic tail causing rapid membrane depolarization and K^+ ion efflux followed by arrest of DNA, RNA and protein synthesis and bacterial cell death. This mechanism depends on Ca^{2+} and phosphatidylglycerol (PG) (Taylor and Palmer 2016). This behaviour has allowed the development of a working model in which the formation of discrete cation-selective pores on lipid bilayers is responsible of the observed permeability to small ions but not to big molecules such as calcein or choline (Chen et al. 2014; Zhang et al. 2014). Indeed, an oligomeric pore formed by four to eight DAP molecules dependently on lipid composition has been recently postulated (Muraih and Palmer 2012; Zhang et al. 2014). Hence, according this model, DAP exerts their membrane perturbation/permeabilization effects through the formation of integral oligomers (Figure 6). Interestingly, lipid acyl chain length has a greater effect on permeabilization than unsaturation or length of the DAP's fatty acid chain, although the presence of oleoyl and other fatty acids could inhibit the pore induction by DAP (Beriashvili et al. 2018). However recent reports show that, instead forming pores in membranes, DAP coclusters with POPG and anionic fluorescent lipid probes, leading to formation of extensive DAP/POPG domains in membranes, causing them to coalesce (Kreutzberger et al. 2017). In any case, DAP also induces local structural changes in lipid bilayers through a so-called 'lipid-extracting' effect which has been revealed using GUV microaspiration and fluorescent microscopy (Chen et al. 2014).

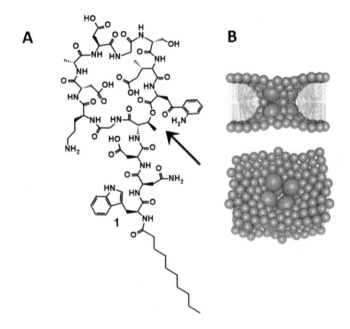

Figure 6. Daptomycin. (A) The common peptide moiety contains 13 amino acid residues, six of which are non-proteinogenic: D-Asn, ornithine (Orn), D-Ala, D-Ser, (2S,3R)-methylglutamate (MeGlu), and kynurenine (Kyn). The C-terminal ten amino acids form a macrocyclic core, which contains a ring-closing depsi (ester) bond between the side chain of Thr and the α-COOH of the C-terminal Kyn (arrow). A N-terminal tripeptide protrudes from the ring and carries the variable fatty acyl residue, which is attached to Trp (position 1). The nature of the fatty acid is a decanoyl residue. (B) Hypothetical model of the membrane-associated DAP octamer that spans both leaflets and forms the functional pore. Adapted from (Taylor and Palmer 2016).

The inhibitory actions of DAP in lipid bilayers, such as is also frequent with other LPs (SF, FG, It-A) are strongly dependent on the lipid composition as well as the critical micellar concentration (*cmc*). Technically, the *cmc* is the *concentration* at which a surfactant (in this case *a LP*) begins to form *micelles*. Hence, the *cmc* can be considered a measure of the efficiency of a LP. In the case of DAP, the *cmc* depends on pH, temperature, and Ca^{2+} and it has a value of 140 µM at pH 3.0, 120 µM at pH 4.0, and 200 µM at pH 2.5 and 5.0 (Qiu and Kirsch 2014). Finally, experiments *in vivo* shows that DAP causes a gradual decrease in membrane potential without forming discrete pores but through rigidification of the lipid membrane, interaction with the inner surface and inhibition of cell wall synthesis (Müller 2016). Notwithstanding the great interest generated by this LP, the complex mechanisms leading to bacterial cell death in DAP-treated microorganisms is still a matter of debate.

2.4. Diversity of Bioactive Lipopeptides

Polymyxin B, lichenysins, and pumilacidins are antimicrobial LPs produced by *B. polymyxa*, *B. licheniformis*, and *B. pumilus* respectively. Overall, as other cyclic LPs, their mode of action is proposed to be membrane disruption leading to a detergent-like activity with strong lipid composition dependence (Hartmann 1978; Meena and Kanwar 2015; Hamley 2015). Serratia *marcescens*, a human pathogen, rod-shaped gram-negative bacteria produces serrawettin W1, a non-ionic LP with biosurfactant properties that was first described as serratamolide. It has been reported to have antimicrobial, antitumor and plant protecting properties as well but how it affects the properties of a lipid bilayer to ions is not clear (Strobel et al. 2005; Diwvedi et al. 2008). Cyclic LPs produced by Pseudomonas spp. also constitute a broad class of chemically diverse biosurfactants divided in 14 distinct groups with multiple structurally homologous members classified as bananamides, pseudofactins, syringomycins, viscosins, orfamides, amphisins, putisolvins, entolysins, xantholysins, tolaasins, fuscopeptins, corpeptins and syringopeptins 22A and 25A (Geudens and Martins 2018). The characterized biological activities of these LPs include processes as diverse as bacterial motility perturbation, antibacterial, antifungal and insecticidal properties, biofilm formation, induced defense responses in plants as well as anti-proliferation effects on human cancer cell-lines. Hence, LPs from this genus are capable of a wide range of biological and functionally relevant effects. However, a complete understanding of the molecular mechanisms through which these compounds exert their activities is lacking. Most of these exotic LPs are able to reduce the surface tension of growth media to different extents with a minimal threshold surface tension of 24.16 mN m^{-1} (Fechtner et al. 2011). On the other hand, studies using the Planar Lipid Bilayer system, unilamellar lipid vesicles, and self-assembled phospholipid monolayers have shown that the syringomycin E (SR-E) forms anion selective voltage-sensitive ion channels that are stabilized by LP molecules of ~1 nm average diameter (Feigin 1996, 1997; Kaulin et al. 2005, Schagina et al. 1998; Malev et al. 2002; Takemoto et al. 2003; Becucci et al. 2015). These ion channels are also sensitive to the membrane dipole potential (Oustramova et al. 2008). Syringopeptins induce larger macroscopic ionic conductances than SR-E but form single channels with similar properties (Bensaci et al. 2011). Viscosin and derivatives such as viscosinamide are cyclic LPs with a highly surface activity synthesized by *Ps. libanensis* or *Ps. fluorescens*. They have antimicrobial activity against plant-pathogenic fungi (*Rhizoctonia solani* and *Pythium ultimum*), capacity to permeabilize or fuse PG/PE/cardiolipin model membrane vesicles and ability to interact both with the polar heads and the aliphatic tails altering bilayer fluidity (Geudens et al. 2017). Finally, a new group of LPs discovered in *Nocardia* sp., peptidolipins, have moderate antibacterial activity against *Staphylococcus aureus* but the mechanism of action underlying these activities is completely unknown (Wyche et al. 2012). In view of all these reports, experimental evidence indicates that membrane perturbation – and in particular pore-formation – is the origin of antimicrobial activity of this diverse group of lipopeptide compounds.

3. Biophysics and Microbial Control

In the context of bio-control of microbial pathogens, lipopeptides have been extensively studied. The three families of *Bacillus* lipopeptides (*i.e.*, surfactin, iturins and fengycins) have diverse antagonistic activities against various fungal phyto-pathogens. In recent past, LPs produced by *Pseudomonas* sp. has also proven effective bio-control agents, specifically against fungal phyto-pathogens as well as they are promising tools in cancer research (Geudens and Martins 2018). Daptomycin is a potent bactericidal of broad spectrum that has been probed its potential for a shorter treatment duration than other antibiotic regimens.

Indeed, in 2003, daptomycin for injection (Cubicin®), was approved by the FDA at a dosage of 4 mg/kg given once daily for the treatment of complicated skin and skin-structure infection (SSSI) caused by specific gram-positive bacteria. In another phase 2 trial, a dosage of 6 mg/kg given in 2 divided doses per day to patients with *S. aureus* bacteremia provided promising results (Eisenstein et al. 2010). Thus, the structural and biological diversity of the cyclic LPs antibiotics has attracted the attention to chemical, agronomical, pharmacological and medical sciences. Future rational drug design, chemical modification and discovery of novel cyclic LPs from natural sources will give rise to useful biological tools for research and development in the biomedical sciences. Indeed, several cyclic lipodepsipeptides and analogues have been chemically synthesized by solid-phase methods using the fluorenyl-methyloxycarbonyl (Fmoc) based chemistry (Etchegaray and Machini 2013) or by mean of genetic approach (Chiocchini et al. 2006).

Thus, the available data derived from an increasingly intense biophysical research of this versatile group of metabolites have open interesting biotechnological perspectives of great potential in order to develop new agronomic and pharmaceutical strategies for the treatment of microbial diseases caused by different pathogens.

CONCLUSION

The interesting physicochemical properties of lipopeptides, and the mechanisms of action through which they exert their antibiotic effects, make these metabolites very interesting biological tools for microbial control. Cyclic lipopeptides are novel and ecologically friend solutions in management of plant diseases, food preservation, and as nontoxic and biodegradable surfactants, as well as they are more and more important in antiparasitic, antiviral, and cancer research.

Particularly useful in agriculture, they are alternative tools to overcome increasing chemical resistance of diverse economically important phytopathogens and with the additional benefit of being highly stable and non-polluting biomolecules. To take advantage of this potential requires detailed molecular biophysical studies to understand the interactions that such molecules establish at level of the lipid bilayer, as well as the understanding of the effect of a large number of variables that affect them. Today, modern biophysics has powerful technical tools capable of helping us solve these molecular puzzles and contribute to implement new bio-control strategies.

ACKNOWLEDGMENTS

The authors gratefully acknowledges the financial support from the Universidad Autónoma de Guadalajara and also thanks Dr. Gloria Macedo for her technical support in obtaining MALDI-TOF mass spectra, as well as the facilities provided by the laboratories of Dr. Miguel J. Beltrán-García (UAG) and Dr. Andrea Alessandrini (UniMore). Finally we would like to thank Professor Juan Villafaña Rojas for providing valuable suggestions to substantially improve the manuscript.

REFERENCES

Abdel-Mawgoud AM, Stephanopoulos G. 2017. Simple glycolipids of microbes: Chemistry, biological activity and metabolic engineering. 2017 *Synth Syst Biotechnol.* 3:3-19.

Andersen OS, Koeppe RE 2nd. 2007. Bilayer thickness and membrane protein function: an energetic perspective. *Annu Rev Biophys Biomol Struct.* 36:107-30.

Ashrafuzzaman M, Tseng CY, Tuszynski JA. 2014. Regulation of channel function due to physical energetic coupling with a lipid bilayer. *Biochem Biophys Res Commun.* 445:463-8.

Balleza D, Alessandrini A, García-Beltrán MJ. 2019. Role of lipid composition, physicochemical interactions, and membrane mechanics in the molecular actions of microbial cyclic lipopeptides. *J Membr Biol.* 252:131-157.

Becucci L, Tramonti V, Fiore A, Fogliano V, Scaloni A, Guidelli R. 2015. Channel-forming activity of syringomycin E in two mercury-supported biomimetic membranes. *Biochim Biophys Acta.* 1848:932-41.

Beltran-Garcia E, Macedo-Raygoza E, Villafana-Rojas J, Martinez-Rodriguez A, Chavez-Castrillon Y. Y, EspinosaEscalante F. M, Mascio P. D, Ogura T, and Beltran-Garcia MJ. 2017. Production of lipopeptides by fermentation processes: Endophytic bacteria, fermentation strategies and easy methods for bacterial selection. Fermentation Processes. InTech. 199-222.

Bensaci MF, Gurnev PA, Bezrukov SM, Takemoto JY. 2011. Fungicidal activities and mechanisms of action of *Pseudomonas syringae* pv. *syringae* lipodepsipeptide Syringopeptins 22A and 25A. Front Microbiol. 2:216.

Beriashvili D, Taylor R, Kralt B, Abu Mazen N, Taylor SD, Palmer M. Mechanistic studies on the effect of membrane lipid acyl chain composition on daptomycin pore formation. *Chem Phys Lipids.* 2018. 216:73-79.

Biniarz P, Łukaszewicz M, Janek T. 2017. Screening concepts, characterization and structural analysis of microbial-derived bioactive lipopeptides: a review. *Crit Rev Biotechnol.* 37:393-410.

Bouffioux O, Berquand A, Eeman M, Paquot M, Dufrêne YF, Brasseur R, Deleu M. 2007. Molecular organization of surfactin-phospholipid monolayers: effect of phospholipid chain length and polar head. *Biochim Biophys Acta.* 1768:1758-68.

Brader G, Compant S, Mitter B, Trognitz F, Sessitsch A. 2014. Metabolic potential of endophytic bacteria. *Curr Opin Biotechnol.* 27:30-7.

Brasseur R, Braun N, El Kirat K, Deleu M, Mingeot-Leclercq MP, Dufrêne YF. The biologically important surfactin lipopeptide induces nanoripples in supported lipid bilayers. *Langmuir.* 2007. 23:9769-72.

Buchoux S, Lai-Kee-Him J, Garnier M, Tsan P, Besson F, Brisson A, Dufourc EJ. 2008. Surfactin-triggered small vesicle formation of negatively charged membranes: a novel membrane-lysis mechanism. *Biophys. J.* 95:3840–3849.

Carrillo, C, JA. Teruel, Aranda A Ortiz. 2003. Molecular mechanism of membrane permeabilization by the peptide antibiotic surfactin. *Biochim. Biophys. Acta.* 1611:91–97.

Chen YF, Sun TL, Sun Y, Huang HW. 2014. Interaction of daptomycin with lipid bilayers: a lipid extracting effect. *Biochemistry.* 53:5384-92.

Chiocchini C, Linne U, Stachelhaus T. 2006. *In vivo* biocombinatorial synthesis of lipopeptides by COM domain-mediated reprogramming of the surfactin biosynthetic complex. *Chem Biol.* 13:899-908.

Cochrane SA, Vederas JC. 2014. Lipopeptides from *Bacillus* and *Paenibacillus* spp.: A gold mine of antibiotic candidates. *Med Res Rev.* 36:4-31.

Deleu M, Lorent J, Lins L, Brasseur R, Braun N, El Kirat K, Nylander T, Dufrêne YF, Mingeot-Leclercq MP. 2013. Effects of surfactin on membrane models displaying lipid phase separation. *Biochim Biophys Acta.*

Deleu M, Paquot M, Jacques P, Thonart P, Adriaensen Y, Dufrêne YF. 1999. Nanometer scale organization of mixed surfactin/phosphatidylcholine monolayers. *Biophys J.* 77:2304-10.

Deleu M, Paquot M, Nylander T. 2005. Fengycin interaction with lipid monolayers at the air-aqueous interface-implications for the effect of fengycin on biological membranes. *J Colloid Interface* Sci. 283:358-65

Deleu M, Paquot M, Nylander T. 2008. Effect of fengycin, a lipopeptide produced by *Bacillus subtilis*, on model biomembranes. *Biophys J.* 94:2667-79.

Dwivedi D, Jansen R, Molinari G, Nimtz M, Johri BN, Wray V. 2008. Antimycobacterial serratamolides and diacyl peptoglucosamine derivatives from *Serratia* sp. *J Nat Prod*. 71:637-41.

Eeman M, Berquand A, Dufrêne YF, Paquot M, Dufour S, Deleu M. Penetration of surfactin into phospholipid monolayers: nanoscale interfacial organization. *Langmuir*. 2006. 22:11337-45.

Eeman M, Deleu M, Paquot M, Thonart P, Dufrêne YF. Nanoscale properties of mixed fengycin/ceramide monolayers explored using atomic force microscopy. *Langmuir*. 2005. 21:2505-11.

Eeman M, Francius G, Dufrêne YF, Nott K, Paquot M, Deleu M. Effect of cholesterol and fatty acids on the molecular interactions of fengycin with *Stratum corneum* mimicking lipid monolayers. *Langmuir*. 2009a. 25:3029-39.

Eeman M, Pegado L, Dufrêne YF, Paquot M, Deleu M. Influence of environmental conditions on the interfacial organisation of fengycin, a bioactive lipopeptide produced by *Bacillus subtilis*. *J Colloid Interface Sci*. 2009b. 329:253-64.

Eisenstein BI, Oleson FB Jr, Baltz RH. 2010. Daptomycin: from the mountain to the clinic, with essential help from Francis Tally, MD. *Clin Infect Dis*. 50 Suppl 1:S10-5.

El Arbi A, Rochex A, Chataigné G, Béchet M, Lecouturier D, Arnauld S, Gharsallah N, Jacques P. 2016. The Tunisian oasis ecosystem is a source of antagonistic *Bacillus* spp. producing diverse antifungal lipopeptides. *Res Microbiol*. 167:46-57.

Etchegaray A, de Castro Bueno C, de Melo IS, Tsai SM, Fiore MF, Silva-Stenico ME, de Moraes LA, Teschke O. 2008. Effect of a highly concentrated lipopeptide extract of *Bacillus subtilis* on fungal and bacterial cells. *Arch Microbiol*. 190:611-22.

Etchegaray A and Machini MT. 2013. Antimicrobial lipopeptides: *in vivo* and *in vitro* synthesis. *In:* Microbial pathogens and strategies for combating them: science, technology and education (A. Méndez-Vilas, Ed.).

Fechtner J, Koza A, Sterpaio PD, Hapca SM, Spiers AJ. 2011. Surfactants expressed by soil pseudomonads alter local soil-water distribution, suggesting a hydrological role for these compounds. *FEMS Microbiol Ecol.* 78:50-8.

Feigin AM, Schagina LV, Takemoto JY, Teeter JH, Brand JG. 1997. The effect of sterols on the sensitivity of membranes to the channel-forming antifungalantibiotic, syringomycin E. *Biochim Biophys Acta.* 1324:102-10.

Feigin AM, Takemoto JY, Wangspa R, Teeter JH, Brand JG. 1996. Properties of voltage-gated ion channels formed by syringomycin E in planar lipid bilayers. *J Membr Biol.* 149:41-7.

Fiechter, A. 1992. Biosurfactants: moving towards industrial application. *Trends Biotechnol.* 10: 208–217.

Francius G, Dufour S, Deleu M, Paquot M, Mingeot-Leclercq MP, Dufrêne YF. 2008. Nanoscale membrane activity of surfactins: influence of geometry, charge and hydrophobicity. *Biochim Biophys Acta.* 1778:2058-68.

Fuertes G, Giménez D, Esteban-Martín S, Sánchez-Muñoz OL, Salgado J. 2011. A lipocentric view of peptide-induced pores. *Eur Biophys J.* 40:399-415.

Geudens N, Martins JC. 2018. Cyclic lipodepsipeptides from *Pseudomonas* spp. - Biological Swiss-Army Knives. *Front Microbiol.* 9:1867.

Geudens N, Nasir MN, Crowet JM, Raaijmakers JM, Fehér K, Coenye T, Martins JC, Lins L, Sinnaeve D, Deleu M. 2017. Membrane interactions of natural cyclic lipodepsipeptides of the viscosin group. *Biochim Biophys Acta.* 1859:331-339.

González-Jaramillo LM, Aranda FJ, Teruel JA, Villegas-Escobar V, Ortiz A. 2017. Antimycotic activity of fengycin C biosurfactant and its interaction with phosphatidylcholine model membranes. *Colloids Surf B Biointerfaces.* 156:114-122.

Goñi FM (2014) The basic structure and dynamics of cell membranes: an update of the Singer-Nicolson model. *Biochim Biophys Acta.* 1838:1467-76.

Gordillo MA, Navarro AR, Maldonado MC. 2015. Mode of action of metabolites from *Bacillus* sp. strain IBA 33 on *Geotrichum citri-aurantii* arthroconidia. *Can J Microbiol.* 61:876-80.

Grau A, Ortiz A, de Godos A, Gómez-Fernández JC. 2000. A biophysical study of the interaction of the lipopeptide antibiotic iturin A with aqueous phospholipid bilayers. *Arch Biochem Biophys.* 377:315-23.

Hamley IW. 2015. Lipopeptides: from self-assembly to bioactivity. *Chem Commun* (Camb). 51:8574-83.

Harnois I, Maget-Dana R, Ptak M. Methylation of the antifungal lipopeptide iturin A modifies its interaction with lipids. *Biochimie.* 1989 71:111-6.

Hartmann W, Galla HJ, Sackmann E. 1978. Polymyxin binding to charged lipid membranes. An example of cooperative lipid-protein interaction. *Biochim Biophys Acta.* 510:124-39.

Heerklotz H, Seelig J. 2001. Detergent-like action of the antibiotic peptide surfactin on lipid membranes, *Biophys. J.* 81:1547–1554.

Heerklotz H, Seelig J. Leakage and lysis of lipid membranes induced by the lipopeptide surfactin. *Eur Biophys J.* 2007. 36:305-14.

Horn JN, Cravens A, Grossfield A. Interactions between fengycin and model bilayers quantified by coarse-grained molecular dynamics. *Biophys J.* 2013. 105:1612-23.

Huang HW, Charron NE. Understanding membrane-active antimicrobial peptides. *Q Rev Biophys.* 2017. 50:e10.

Kanafani ZA, Corey GR. 2007. Daptomycin: a rapidly bactericidal lipopeptide for the treatment of Gram-positive infections. *Expert Rev Anti Infect Ther.* 5:177-84.

Kaulin YA, Takemoto JY, Schagina LV, Ostroumova OS, Wangspa R, Teeter JH, Brand JG. 2005. Sphingolipids influence the sensitivity of lipid bilayers to fungicide, syringomycin E. *J Bioenerg Biomembr.* 37:339-48.

Koglin A, Doetsch V, Bernhard F. 2010. Molecular engineering aspects for the production of new and modified biosurfactants. *Adv Exp Med Biol.* 672:158-69.

Kolter R. 2010. Biofilms in lab and nature: a molecular geneticist's voyage to microbial ecology. *Int Microbiol.* 2010 Mar;13:1-7.

Kreutzberger MA, Pokorny A, Almeida PF. Daptomycin-Phosphatidylglycerol domains in lipid membranes. *Langmuir.* 2017. 33:13669-13679.

Kung C (2005) A possible unifying principle for mechanosensation. Nature. 436:647-654.

Landy M, Warren GH, Rosenman SB, Colio LG. 1948. Bacillomycin; an antibiotic from *Bacillus subtilis* active against pathogenic fungi. *Proc Soc Exp Biol Med.* 67:539-41.

Lee MT, Hung WC, Chen FY, Huang HW. 2008. Mechanism and kinetics of pore formation in membranes by water-soluble amphipathic peptides. *Proc Natl Acad Sci USA.* 105:5087-92.

Lee MT, Yang PY, Charron NE, Hsieh MH, Chang YY, Huang HW. 2018. Comparison of the effects of Daptomycin on bacterial and model membranes. *Biochemistry.* 57:5629-5639.

Maget-Dana R, Ptak M, Peypoux F, Michel G. Pore-forming properties of iturin A, a lipopeptide antibiotic. *Biochim Biophys Acta.* 1985. 815:405-9.

Malev VV, Schagina LV, Gurnev PA, Takemoto JY, Nestorovich EM, Bezrukov SM. 2002. Syringomycin E channel: a lipidic pore stabilized by lipopeptide? *Biophys J.* 82:1985-94.

Marin-Medina N, Ramírez DA, Trier S, Leidy C. Mechanical properties that influence antimicrobial peptide activity in lipid membranes. *Appl Microbiol Biotechnol.* 2016, 100:10251-10263.

Marquette A, Bechinger B (2018) Biophysical investigations elucidating the mechanisms of action of antimicrobial peptides and their synergism. *Biomolecules.* 8(2). pii: E18.

McIntosh TJ, Simon SA (2006) Roles of bilayer material properties in function and distribution of membrane proteins. *Annu Rev Biophys Biomol Struct.* 35:177-98.

Meena KR, Kanwar SS (2015) Lipopeptides as the antifungal and antibacterial agents: Applications in food safety and therapeutics. *BioMed Research International.* Volume 2015, Article ID 473050.

Mnif I, Ghribi D (2015) Lipopeptides biosurfactants: Mean classes and new insights for industrial, biomedical, and environmental applications. *Biopolymers.* 104:129-47.

Moravej H, Moravej Z, Yazdanparast M, Heiat M, Mirhosseini A, Moosazadeh Moghaddam M, Mirnejad R. 2018. Antimicrobial Peptides: Features, Action, and Their Resistance Mechanisms in Bacteria. *Microb Drug Resist.* 24:747-767.

Müller A, Wenzel M, Strahl H, Grein F, Saaki TN, Kohl B, Siersma T, Bandow JE, Sahl HG, Schneider T, Hamoen LW. 2016. Daptomycin inhibits cell envelope synthesis by interfering with fluid membrane microdomains. *Proc Natl Acad Sci USA.* 113:E7077-E7086.

Mulligan CN. Environmental applications for biosurfactants. *Environ Pollut.* 2005, 133: 183–198.

Muraih JK, Palmer M. Estimation of the subunit stoichiometry of the membrane-associated daptomycin oligomer by FRET. 2012. *Biochim Biophys Acta* 1818:1642-1647.

Neu TR. Significance of bacterial surface-active compounds in interaction of bacteria with interfaces. *Microbiol Rev* 1996, 60: 151–166.

Ostroumova OS, Malev VV, Bessonov AN, Takemoto JY, Schagina LV. Altering the activity of syringomycin E via the membrane dipole potential. *Langmuir.* 2008. 24:2987-91.

Ostroumova OS, Malev VV, Ilin MG, Schagina LV. 2010. Surfactin activity depends on the membrane dipole potential, *Langmuir* 26:15092–15097.

Perez KJ, Viana JD, Lopes FC, Pereira JQ, Dos Santos DM, Oliveira JS, Velho RV, Crispim SM, Nicoli JR, Brandelli A, Nardi RM (2017) *Bacillus* spp. Isolated from Puba as a source of biosurfactants and antimicrobial lipopeptides. Front Microbiol. 8:61.

Phillips R, Ursell T, Wiggins P, Sens P (2009) Emerging roles for lipids in shaping membrane protein function. *Nature.* 459:379-385.

Qiu J, Kirsch LE. 2001. Evaluation of lipopeptide (daptomycin) aggregation using fluorescence, light scattering, and nuclear magnetic resonance spectroscopy. *J Pharm Sci.* 2014. 103:853-61.

Razafindralambo H, Popineau Y, Deleu M, Hbid C, Jacques P, Tonart P, Paquot M (1997) Surface-active properties of surfactin/iturin A mixtures produced by *Bacillus subtilis. Langmuir* 13:6026–6031.

Ron EZ, Rosenberg E. 2001. Natural roles of biosurfactants. *Environ Microbiol.* 3:229-36.

Schagina LV, Kaulin YA, Feigin AM, Takemoto JY, Brand JG, Malev VV. 1998. Properties of ionic channels formed by the antibiotic syringomycin E in lipid bilayers: dependence on the electrolyte concentration in the bathing solution. *Membr Cell Biol.* 12:537-55.

Singh M, Kumar A, Singh R, Pandey KD. Endophytic bacteria: a new source of bioactive compounds. *3 Biotech.* 2017 Oct;7(5):315.

Steenbergen JN, Alder J, Thorne GM, Tally FP. 2005. Daptomycin: a lipopeptide antibiotic for the treatment of serious Gram-positive infections. *J Antimicrob Chemother.* 55:283-8.

Strobel GA, Morrison SL, Cassella M, 2005. Protecting plants from oomycete pathogens by treatment with compositions containing serratamolide andoocydin a from *Serratia marcescens.* Patent Number: US2003049230-A1;US6926892-B2.

Takemoto JY, Brand JG, Kaulin YA, Malev VV, Schagina LV, Blasko K. 2003. The syringomycins. In: Menestrina G, Dalla Serra M and Lazarovici P (ed) Pore-forming peptides and protein toxins, 1st. edn. Taylor & Francis, London. pp 260-271.

Taylor SD, Palmer M. 2016. The action mechanism of daptomycin. *Bioorg Med Chem.* 24:6253-68.

Teng J, Loukin S, Anishkin A, Kung C. 2015. The force-from-lipid (FFL) principle of mechanosensitivity, at large and in elements. *Pflügers Arch.* 467:27-37.

Thimon L, Peypoux F, Wallach J, Michel G. 1995. Effect of the lipopeptide antibiotic, iturin A, on morphology and membrane ultrastructure of yeast cells. *FEMS Microbiol Lett.* 128:101-6.

Vanittanakom N, Loeffler W, Koch U, Jung G. 1986. Fengycin: a novel antifungal lipopeptide antibiotic produced by *Bacillus subtilis* F-29-3. *J Antibiot* (Tokyo). 39:888-901.

Wani ZA, Ashraf N, Mohiuddin T, Riyaz-Ul-Hassan S. 2015. Plant-endophyte symbiosis, an ecological perspective. *Appl Microbiol Biotechnol.* 2015. 99:2955-65.

Wyche TP, Hou Y, Vazquez-Rivera E, Braun D, Bugni TS. 2012. Peptidolipins B-F, antibacterial lipopeptides from an ascidian-derived *Nocardia* sp. *J Nat Prod.* 75:735-40.

Zhang T, Muraih JK, Tishbi N, Herkowitz J, Victor RL, Silverman J, Uwumarenogie SD, Taylor SD, Palmer M, Mintzer E. 2014. Cardiolipin prevents membrane translocation and permeabilization by daptomycin. *J Biol Chem.* 289:11584-1159.

In: Lipid Bilayers
Editor: Mohammad Ashrafuzzaman © 2019 Nova Science Publishers, Inc.
ISBN: 978-1-53616-392-6

Chapter 5

COLCHICINE INDUCED ION CHANNEL FORMATION INTO MEMBRANES AS A MECHANISM BEHIND CHEMOTHERAPY DRUG CYTOTOXICITY OF CANCER CELLS

A. A. Alqarni, S. Zargar and Md. Ashrafuzzaman[*]

Department of Biochemistry, College of Science,
King Saud University, Riyadh, Saudi Arabia

ABSTRACT

We report chemotherapy drug (CD) induced two sets of toxicity results: (i) induction of ion pores into lipid bilayer membranes in an aqueous phase, (ii) cytotoxicity on cancer cell lines. Both of these toxicities happened due to the effects of physiologically relevant CD concentrations in membrane aqueous or cellular environment. Ion pore formation was investigated by CDs: thiocolchicoside (TCC) and taxol (TXL) which primarily target tubulin but not only. TCC interacts with $GABA_A$, nuclear envelope and strychnine-sensitive glycine receptors. TXL interferes with

---

[*] Corresponding Author's Email: mashrafuzzaman@ksu.edu.sa.

the normal breakdown of microtubules and interacts with mitochondria and is found as a significant chemotherapeutic agent for breast, ovarian and lung cancers. CDs were examined for understanding their effects on phospholipid bilayer membranes. Our electrophysiology recordings indicate that CDs induce considerably stable ion flowing pores/channels in membranes. The discrete current versus time plots exhibit triangular shapes which is consistent with a spontaneous time-dependent change of the pore conductance in contrast to rectangular conductance events usually induced by ion channels. The CD induced events exhibit conductance ($\sim$0.01-0.1 pA/mV) and lifetimes ($\sim$5-30 ms) within the ranges observed in peptide-induced events e.g., those of gramicidin A and alamethicin channels. The channel formation probability increases linearly with CD concentration and transmembrane applied potential and is not affected by pH (5.7 - 8.4). A theoretical explanation on the causes of CD induced ion pore formation and the pore stability has been found using our recently discovered binding energy between the lipid bilayer and the bilayer embedded ion channels where we have used gramicidin A channels as tools. This energetics picture suggests that as the channel forming agents approach to the lipids on bilayer the localized polarized charge profiles in the constituents of both channel forming agents (e.g., CDs in this study) and the lipids determine the electrostatic drug-lipid coupling energy through their screened Coulomb interactions. The strength of this electrostatic energy originated coupling determines the stability/lifetime of the drug-induced ion pores following standard Statistical Mechanical formula. Induction of such stable pores may lead to causing considerable cytotoxicity. In the cancer cell line viability assays we tested the colchicine induced cytotoxicity. We found that considerable effects on cell viability happens as a result of the effects of physiologically relevant concentrations of colchicine on breast cancer cell lines. The CD induced severe cancer cell cytotoxicity may be due to various reasons. Our discovered membrane pore formation may account as one of the various molecular mechanisms responsible for CD induced cytotoxicity. These findings may elucidate cytotoxic effects of chemotherapy drugs, raise understanding of a possible molecular mechanism behind CD induced cytotoxicity, and aid in the development of novel drugs for a broad spectrum of cancers and other diseases.

Keywords: chemotherapy drugs, colchicine, lipid bilayer membrane, cancer cell cytotoxicity, ion pore

1. INTRODUCTION

Drug induced cell membrane permeabilization is an important area of research. Electrophysiology recordings of ion currents flowing across membranes doped with various naturally occurring or synthetic membrane proteins (MPs) or antimicrobial peptides (AMPs) may be used to investigate whether a protein-lipid complex gets created and as a result an ion flowing pore/channel across a membrane may be constructed. Using the same technique we have investigated whether chemotherapy drugs (CDs) induce any physical event inside membrane in analogy to those due to AMP-induced ion channels. To investigate this possibility, we have chosen two important CDs colchicine and taxol, as examples, from a list of several hundred chemotherapy drugs and drug candidates that are either in clinical use for patients, in clinical trials or under investigations, see ref [1]. To understand the molecular mechanism of drug effects on membranes we performed molecular dynamics (MD) simulations of various CD-lipid complexes.

Colchicine has a history of use in immune-system diseases [2, 3]. In 2009, colchicine won the USA *Food and Drug Administration* (FDA) approval as a drug for acute gout and familial Mediterranean fever. It is found to inhibit leukocyte-endothelial cell adhesion [4] and T-cells activation [5] by binding to tubulin dimers. This prevents tubulin polymerization into microtubules (MTs) [6]. Due to increased rate of mitosis cancer cells are found to be more vulnerable to colchicine effects than normal cells. Colchicine's therapeutic potential in cancer chemotherapy is quite limited due to its high toxicity against the normal cells. Colchicine is found to shift the dynamic equilibrium in MT's towards disassembly through sequestering most of the tubulins [7] and inducing slow disassembly of the MT's [8]. This may account for gradual change of the membrane action potential (towards decrease) or threshold (towards increase) and the resting potential, found to be altered modestly at the colchicine concentrations of ~2.5 mM. At higher concentrations (~10 mM) the resting potential was reduced by up to as high as 5 mV [7]. Colchicine is also detected to bind with nuclear periphery, and to create disorders in the nuclear membrane phospholipid bilayers [9]. Since the FDA's approval in 1992 taxol has rigorously been used for ovarian, lung and breast cancer chemotherapy [10], due to its molecular action involving stabilization of the MTs [11], making it especially quite suitable in combination therapy [12]. However, the issue of taxol's poor solubility requires conjugation with cremophor or albumin. A taxol-phospholipid liposome construct may increase taxol's antitumor efficacy by delaying the tumor progression compared to the case where free taxol is administered in cremophor [13]. Taxol to found to inhibit endosomal-lysosomal membrane trafficking by favorably inducing small condensed vesicles over the large ones [14]. This suggests that taxol affects the lipid curvature profiles. Taxol incorporated into liposomes, found to penetrate into the acyl chain domain which alters the physical properties e.g., the phospholipid phase transitions, the lipid order parameters, the fluidity, etc. [15] of both of the artificial and the biological membranes. Both MT-stabilizing taxol and MT-disrupting colchicine were found not to affect human platelets' membrane lipid fluidity [16]. The strong interactions of colchicine with lipids [17] suggests that it is a poorer candidate as an anti-cancer drug due to lipid adsorption, by being weakly accessible to tubulin. Although the activity of colchicine and taxol has been found to be regulated due to lipids or liposomes, there is still lack of concrete information about the effects of these molecules on cell membranes. Little is known about colchicine's effects on membrane transport properties that get influences by physical, geometrical and structural characteristics of membranes. Understanding these effects will shed light on off-target interactions of these chemotherapeutic agents, and hence may lead to discovering an improved drug design template.

CDs colchicine and taxol have proven effects on membrane's certain electrical and physical properties [7, 9, 14-17]. The understanding of their general effects on membrane's transport properties and the specific drug/membrane constituent interactions is quite important from the cytotoxicity point of view, but remains largely unexplored. Properties like relatively small in size and apparently neutral in charge effects of these drug molecules raise the possibility that they may act differently on membranes than most of the antimicrobial agents. The latter group members are rigorously reported to undergo considerable conformational changes as they cross through the hydrophilic/hydrophobic boundary while entering into the lipid membrane environment (e.g., see refs. [18, 19]), and some of them cause membrane general permeabilization through formation of the well-structured long-lived protein-lined [20-22] ion channels. Their amphiphile type effects would otherwise modulate at least the bilayer's physical properties to some extent. Comparable effects of amphiphiles such as triton X-100 and capsaicin [23] and small size AMP gramicidin S (GS) at nanomolar (nM) concentrations [24] on bilayer mechanical properties, or transient disordering defects induced by GS at micromolar (μM) concentrations in the lipid layers [25] were observed.

The goal of ongoing research was two fold. First goal was to investigate CD effects on model membrane systems. Second goal was to inspect the gross cytotoxicity on cell line as a result of the effects of colchicine.

Firstly, to evaluate the two specific tubulin binding drug molecules thiocolchicoside (TCC), a derivative of colchicine, and taxol (TXL) (see Figure S1 for their structures) regarding their effects on model lipid membranes using electrophysiology techniques. These molecules not only permeabilize lipid bilayer membranes, but also are found to create certain molecular structures through reorganization among lipid molecules and themselves. Evidence is found that they form stable ion pores [26-28] with unconventional characteristics inside model lipid membranes involving the CDs. MD simulation results provide computational support for the drug-lipid physical interactions which are postulated to be the mechanisms behind such pore formation. An attempt to better understand the drug-lipid interactions will be made based on screened Coulomb and van der Waals (vdW) interactions that were recently reported to be the main mechanisms behind the creation of ion channels, especially investigated for AMPs gramicidin A (gA) and alamethicin (Alm) [29-32].

After elucidating the ion pore creation by CDs we aimed at addressing their general cytotoxicity. In this second phase, we elaborated our investigations by seeing the effects of various concentrations of colchicine on cell lines.

For initial screening, we chose a common cancer cell line, which is breast cancer MCF-7 to study the cytotoxicity assay and check the sensitivity of the cells to CD cytotoxic effect. Breast cancer remains one of the most common cancers affecting and disturbing women universally [33] with up to 30% of all breast cancers estimated to ultimately metastasize [34]. Cytotoxicity testing was completed using a common cell viability assay, the MTT assay. A primary MTT assay is generally used to test the number of cells ideal for use in the cytotoxicity assay [35]. This assay is considered reproducible *in vitro* test methods for screening potentially many toxic agents (National Institutes of Health, 2006)).

The cytotoxicity assays in cell cultures may be used to advantage and benefit for assessing the general toxicological potential of phototherapeutics samples [36]. Mossman in 1983 described for the first time that the assay depends on reduction of MTT tetrazolium salt to formazan that take place in the mitochondria of living cells, due to the activity of mitochondrial dehydrogenases. Lidia et al. found that the compounds that modify cell metabolism by increasing the NADPH level or the activity of LDH (lactate dehydrogenase) significantly get influenced by MTT assay [37]. Sonia et al. has reported that 1-methyl-5-(3-(3,4,5-trimethoxyphenyl)-4H-1,2,4-triazole-4-yl)-1H-indole designated T115 is found to inhibit tubulin polymerization by targeting the colchicine-binding site in β-tubulin protein [38]. T115 showed a dose-dependent inhibitory effect on the polymerization of tubulin heterodimers, suggesting that T115 is a potent microtubule polymerization inhibitor. This group has shown that T115 competes with colchicine for its binding pocket in tubulin structure, produces robust inhibition of tubulin polymerization, and disrupts the microtubule. Luciana et al. studied the cytotoxicity of plant extract assessed by MTT and neutral red (NR) assays. Aqueous extract added to the culture medium presented the best profile to assess cytotoxicity [39]. Considering the efficacy of the technique and compounds as explained here, we considered the MTT assay for addressing our cytotoxicity of colchicine to see what in gross may happen on cell lines that may originate partly due to the molecular mechanisms (ion pore formation) we have reported here.

2. MATERIALS AND METHODS

2.1. Electrophysiology Experiments

Lipid bilayers were constructed by applying a cocktail of lipids Phosphoethanolamine: phosphatydyleserine: phosphatidylcholine (5:3:2, v/v/v)/n-decane using the painting method over a cross sectional 150 μm septum of a bilayer cuvette.

The volume of cuvette was 1 mL containing buffer (0.5 M NaCl, 10 mM HEPES, pH 7.4) in both *cis* (recording electrode site) and *trans* (reference electrode site) chambers. Stocks of TCC (purchased from ChemRoutes, Canada) or TXL (purchased from Vector Lab, Canada) were prepared in dissolving Dimethylsulfoxide (DMSO) (4 mg/mL or 7.1 mM), then diluted in buffer for experimental use (1 mg/mL). Following a vigorous vortexing the stock of TCC or TXL was added to the *cis* chamber buffer while stirring to avoid any possible drug solubility issues. HEPES was replaced with 2-(N-morpholino) ethanesulfonic acid (MES) and TAPS (2-hydroxy-1,1-bis(hydroxymethyl)ethyl)amino-1-1 propanesulfonic acid)) in buffer for solutions with different pHs 5.7 and 8.4, respectively.

Following the formation of bilayers, we waited for about 1 hr, then tested the bilayer stability by applying transmembrane potential of V=±400 mV for 3 minutes. After the addition of drugs we waited for 20 minutes before recording the currents. The ion pore activities at reasonable values of V were recorded as current traces. Experiment was repeated in three independent conditions. Ussing Chamber Systems (from Physiologic Instruments, San Diego, CA 92128) were used for this study of the transport of membranes. For data acquisition Clampex 8.2 (from Axon Instruments, CA, USA) was used. pA order currents through membranes were recorded at the filter frequency 20 kHz. The current was then plotted as a function of recorded time using software Origin 8.5 (from OriginLab Corp., Northampton, MA).

2.2. MD Simulation

The *in silico* CD-lipid interactions were modeled in MD methodology. Based on the Monte Carlo concept, we considered five different relative locations and orientations randomly generated in each CD-lipid complex created by a CD and a lipid molecule as initial structures for MD simulations to increase sampling size for better statistical analysis. For each location- and orientation-specific complex, a 6 ns explicit water MD simulation at a temperature of 300K in an aqueous solution at pH 7 was performed. We applied the software package Amber 11 [39, 40], specifically the Amber force field ff03 was used. The explicit water TIP3P model was used to simulate solvent effects. The force field parameters for CDs and lipids (PC and PS) were generated using an Amber module antechamber [41, 42]. Note that the force field parameters for CDs are similar to the ones generated for colchicine [43] and taxol [44] respectively. Both studies [43, 44] have shown these parametrizations lead to simulations to be consistent with experiments. Therefore, we can expect to observe similar results based on these parametrizations in current studies. Twenty complexes were energy minimized using the steepest descent method for the first ten cycles and then followed by a conjugate gradient for another 1000 cycles. We then applied Langevin dynamics during the process of heating up the system for 200 ps with the energy minimized complex, in which drug and lipid molecules were being restrained using a harmonic potential with a force constant k=100 N/m. Afterward, we introduced pressure regulation to equilibrate water molecules around the complex and to reach an equilibrium density for another 200 ps in addition to temperature regulation. The MD production run then was continued for 6 ns. Given five various initial structures for each lipid-drug pair, a total 30 ns simulation result was analyzed to gain insights of the direct interactions of the corresponding pair. Note that the phospholipid was gently restrained with a harmonic potential with a force constant k=10 N/m, applied only to the phosphate during the production runs. The purpose of this restraint is to mimic a single phospholipid being "restrained" in membrane while both head group and two tails of such lipid still possess certain degrees of freedom.

2.3. Colchicine Cytotoxicity assay on Breast Cancer Cell Lines

2.3.1. Materials

96 wells plate (Becton Dickinson Labware), DMEM media (Invitrogen), breast cancer cells (ATCC), Trypsin (Invitrogen), EDTA (Sigma), MTT staining solution (National scientific company), Formazan dilute solution (National scientific company), FCS buffer (Invitrogen), 5% CO_2 incubator (Memmert), PBS buffer (Amblicon), Colchicine drug (Sigma), Micropipettes, at -20°C to stored MTT solution(National scientific company) and 4°C to stored Formazan solution (National scientific company), 570 nm spectrophotometer (Amarsham) and 595nm ELISA reader (Biotek Elx-100).

2.3.2. Colchicine Cytotoxicity Assay Protocol by Using MTT Assay Method

MTT assay was done to check the viability of cells before and after treatments. The breast cancer cells (MCF-7) were passaged as 100 μl cells per well in 96-Well culture plates in duplicate. 1000 μl DMEM media was added to each well of cells and incubated at 37°C with 5% CO_2 for 4 hours to let them adhere to the surface (cells number for each well was similar).

Table 1. Concentration gradient of colchicine drug

100 μl cells	100μl cells	100 μl cells + 1 μl drug	100 μl cells + 1 μl drug	100 μl cells + 2 μl drug	100 μl cells + 2 μl drug
100 μl cells + 4 μl drug	100 μl cells + 4 μl drug	100 μl cells + 6 μl drug	100 μl cells + 6 μl drug	100 μl cells + 8 μl drug	100 μl cells + 8 μl drug
100 μl cells + 10 μl drug	100 μl cells + 10 μl drug	100 μl cells + 12 μl drug	100 μl cells + 12 μl drug	Media + 10μl MTT + 100 μl formazan	Media + 10μl MTT + 100 μl formazan

Later concentration gradient of colchicine drug (experimental stock: 0.003 mg Col. + 990 μl media + 10 μl DMSO) was added to each well of adherent cells except control. 0-10 μl concentration gradient in duplicate was added to wells. Cells were incubated with drug for 24 hours in 5% CO_2 incubator at 37°C. Cells were observed under the microscope to check the behavior with treated drug concentrations. Drug concentrations were as follows:

After 24 hours 10 µl MTT reagent was added to each well and cells were continued to culture for another 4 hours. Later 100µl formazan was added to each well and plate was shaken at low speed for 10 min until crystals, if any, had dissolved completely (we do not need this step in our experiments due that there was no crystal formed when we added formazan chemical). Then the absorbance was measured at 570 nm using spectrophotometer and at 595 nm using ELISA reader against blank. Blank well was having (culture-media, MTT staining solution, formazan solution) and other wells (comparable well) (cells, drug dissolve media with same concentration, nutrient medium, MTT staining solution, formazan solution).

3. RESULTS

Before presenting the drug cytotoxicity results we present two other sets of results, found *in vitro* and *in silico* experiments demonstrating the membrane and/or membrane constituents binding based drug effects. Firstly, we present the results concerning the changes of the lipid membrane's electrical conductance properties due to the effects of CDs TCC and TXL. Secondly, we describe MD results which illustrate computational predictions regarding drug-lipid interactions between drug molecules and lipid membranes once they find themselves in each other's close proximity.

Electrophysiology results presented in Figure 1 (long time traces are also presented in Figure S2.A) show considerable conductance events in membranes doped with TCC or TXL [45]. An intact lipid membrane in aqueous phases is generally non-conducting to ions (see Figure S2.B) but in the presence of TCC/TXL different current levels were observed. Long-time records indicate that time independent/random appearance of current events was induced by both compounds. The stability of current levels varied significantly. The point count plot (presented in Figure S2.A) shows that the conductance events do not correspond to discrete current levels with discrete conductance values. These current levels appeared with various possible conductance values, similarly to events observed due to GS [25]. Although no major qualitative difference between TCC and TXL-induced membrane permeabilization was observed, TXL-induced current levels covered a relatively lower conductance range than those for TCC. It is also worth mentioning that after breaking a membrane doped with drugs and recording the current traces on a reconstructed membrane we observed a reduced appearance of current events (traces not shown here and were not used in analysis).

In Figure 1, short-time records show independent conductance events. Any current event or conductance burst (see Figure S2.A) can be due to an independent drug induced conductance event or a combination thereof. We observed no current events across lipid membranes without being doped with drugs. To demonstrate that we have presented a 60 s recorded current trace in Figure S2.B. A current event does not represent any control or a reproducible parameter which could elucidate the nature of the conducting pore induced by a drug. Note that the independent conductance events appear with triangular shapes which means that the conductance in a single event is not constant but increases/decreases spontaneously over the time interval comparable to the 'lifetime' of any specific conductance event. To the best of our knowledge, the appearance of triangular conductance events with time dependent change of conductance is a new and unique phenomenon. The amplitudes of these events are also different from typical values. We observed random spontaneous transitions between different current levels within a discrete conductance event during its lifetime. These discrete events were found to be approximately characterized by conductance values in the range ~0.01-0.1 pA/mV and lifetimes in the range ~5-30 ms based on randomly chosen 20 discrete conductance events. The lowest possible value for both conductance and lifetime should ideally approach 0 but events with lower than the above-mentioned values are masked by the noise. For comparison, the conductance events of gA and Alm channels are also shown. Similarly to the conductance events in

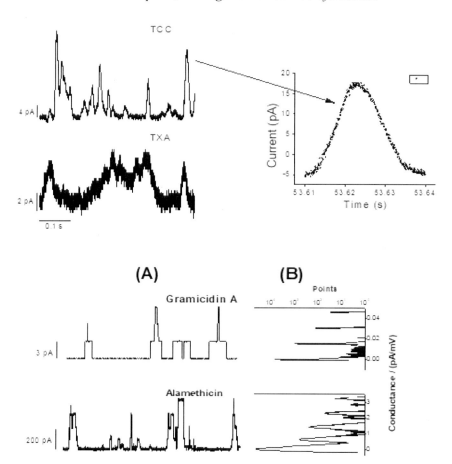

Figure 1. The upper panel shows triangular-shape conductance events induced by TCC and TXL, both at 90 μM. pH=5.7, V=100 mV. Both traces were filtered at 20 kHz but the lower one shows higher noise due to its presentation (the current axis) at an amplified scale. In a high resolution plot (shown on the right-hand side of the arrow) of a single event only showing individual points (in Origin 8.5 plot) we observe all points (open circle) with increasing/decreasingcorresponding values of conductance, respectively, at both left and right lateral sides of the chemotherapy drug induced triangular conductance events. The lower panel (A) illustrates rectangular-shape conductance events in gA and Alm channels (see e.g., [25]). gA channel activity was recorded at 200 mV and Alm at 150 mV. Traces representing gA and Alm channel activities in phospholipid bilayers were recorded at filter frequencies 2 kHz and 20 kHz, respectively. A lower filter frequency for traces representing gA channel activity is acceptable because of the channel's relatively higher stability. In (B) the point count plots of the current traces through gA and Alm channels the peaks are found at discrete values of conductance.

channels formed by integral membrane proteins or antimicrobial peptides, both gA and Alm events have rectangular shapes. This means that transitions between discrete current levels are transient or extremely abrupt. In gA channels only a back-and-forth transition exists between two current levels representing dimer/monomer states. In Alm channels, there exist many transitions between different conductance states consisting of various numbers of monomers. All transitions are transient that means the current fluctuations at the transition take almost no time. A time-dependent current fluctuation between random current levels in any TCC/TXL-induced conductance event is therefore an important novel finding reported here.

There is an important feature observed in the CD-induced triangular conductance events which is that the time for the conductance to rise and fall for each individual event, is rather consistent. From many such events the rise and fall times can be used to characterize the conductance. These times may be influenced by the applied voltage or by the concentration of drug present and thus can manifest the kinetics of channel growth and decay. The time dependent increase and decrease of the pore conductance which can be considered as pore conductance changes over time were therefore analyzed under experimental conditions stated in Figure 1 and found to be 1.61±0.46 pA/mV.s and 1.71±0.47 pA/mV.s for TCC induced pore and 0.76±0.27 pA/mV.s and 0.75±0.23 pA/mV.s for TXL induced pore, respectively. These values were derived by investigating 20 randomly chosen discrete conductance events for both TCC and TXL. The corresponding values for peptide induced pores/channels (e.g., gA and Alm channels) approach to infinity as in both cases the transitions between different conductance levels take no measurable time.

A linear dependence was observed for values of pore activity (A) (below 1.0) on V and TCC/TXL concentration (c_D) for up to very high values of both (see Figure S3). Pore activity can be defined as the duration of time when the membrane was found to be permeable to ions relative to the total current record time. It is not feasible to state lower/upper cut off values for V and c_D due to their interdependence. We also observed conductance events (data not shown) at a much lower range and for more biologically relevant (to cancer therapy) concentrations e.g., at c_D=10 µM for both TCC and TXL but at V~200 mV or higher. Although we rigorously vortexed and then stirred the stock while adding into the aqueous phase we can not rule out the possibility of solubility issues. We therefore predict that by overcoming the possible solubility issues of these drug molecules with concomitant use of other solubility enhancing agents (which is a possible choice in therapy) it should be possible to observe membrane conductance events induced by TCC and TXL at even lower concentrations which are physiologically relevant in the context of cancer treatment.

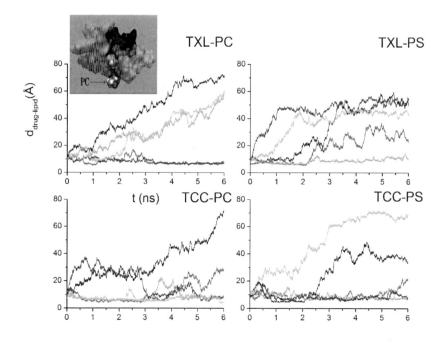

Figure 2. MD simulation results representing the change of the CD-lipid center of mass distance $d_{\text{drug-lipid}}$ with time. Five curves with different colors represent five independent initial CD-lipid complexes. The inset (only TXL-PC is shown) shows the cartoon representations of initial structures of five complexes with TXL following the color of the corresponding curve.

No considerable change of pore activity in the range of pH values between 5.7 and 8.5 was observed (see Figure S4).

After observing the effects of drugs on membrane's conductance we performed MD simulations to investigate CD-lipid interactions in molecular level. Figure 2 shows the MD results using five CD-lipid complexes as initial structures as shown in the inset. It plots the separation distance of centers of mass of CD and lipid molecules, $d_{\text{drug-lipid}}$, against simulation time t (ns). Note that $d_{\text{drug-lipid}}$ was used as the simplest property to quantify the effects of CD-lipid interactions. It shows that $d_{\text{drug-lipid}}$ fluctuates around 10 Å, similarly to the initial setting in two to three simulations in TXL-PC, TCC-PC and TXL-PS, TCC-PS. Drugs and lipids were observed to be gradually separated in most of the simulations. Figure 3 plots snapshots of CD-lipid complex at the beginning (left) and at 6th ns (right) of the simulations. Note that it only presents the case that a CD likely to bind with the single lipid. Although TXL in both blue and red cases shown in Figure 2 starts with different orientations, the simulations indicate TXL likely to bind to similar location that near phosphate group domain shown in Figure 3 (A). We also observe TCC (cyan case in Figure 2) tends to bind to similar location (Figure 3 (B)). Yet Figure 3 (A) and (B) shows TXL and TCC (green case in Figure 2) likely insert into the cavity between two short tails of PS respectively.

The solvent accessible (SA) area of the complex in the MD simulations was used to investigate whether the hydrophobic effects contribute to CD-lipid binding. Figure 4 shows SA areas in all four cases against $d_{\text{drug-lipid}}$. When both drug and lipid molecules are completely separated we can expect them to be entirely exposed to solvent, i.e., the corresponding SA areas are at a maximum. The figure shows that the SA areas in all four cases are unchanged between the start and the investigated 20 Å length. This suggests that within this range the drug lipid complexes stay at an equilibrium solvation condition.

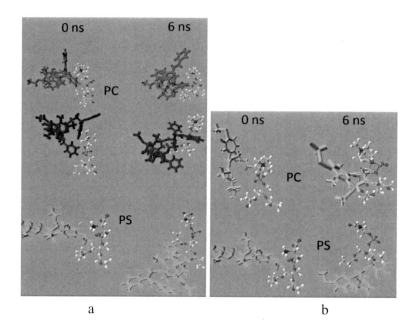

Figure 3. Snapshots of MD simulations. CD-lipid complex at the beginning (left) and 6th ns (right) of the simulation. The colorings of CDs are corresponded to the color definitions used in in Figure 2. (A) It shows structures of TXL-PC (upper and mid panel) and -PS (bottom); (B) it shows structures of TCC-PC (top) and –PS (bottom).

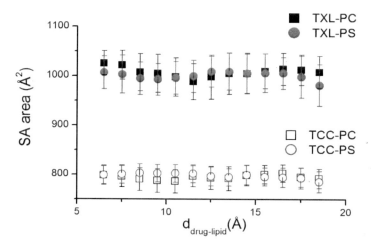

Figure 4. Solvent accessible (SA) areas for four complexes are plotted against the CD-lipid center of mass distance $d_{drug-lipid}$. It shows SA areas are independent of $d_{drug-lipid}$ for four complexes.

Colchicine Induced Ion Channel Formation into Membranes ... 167

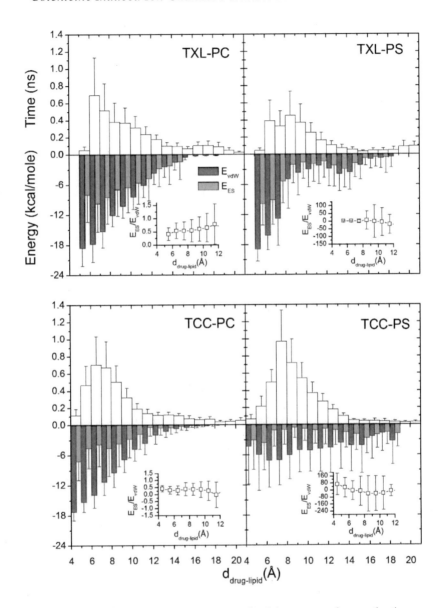

Figure 5. In all four histogram plots (upper panel) of time versus $d_{\text{drug-lipid}}$ the time durations when drug/lipid stay together (height) within a distance (width) during 6 ns simulations are presented. Lower panels show the histograms of non-bonded van der Waal's (vdW) energy (E_{vdW}) and electrostatic (ES) interactions energy (E_{ES}). To avoid color conflict E_{vdW} and E_{ES} are shown to occupy half-half widths though each half represents the whole width of the corresponding histogram.

Histograms of $d_{drug-lipid}$ from all five 6 ns simulations and the corresponding energy contributions from two non-bonded interactions, van der Waals (E_{vdW}) and electrostatic forces (E_{ES}) versus $d_{drug-lipid}$ are shown in Figure 5. The histogram of $d_{drug-lipid}$ shows that both TXL and TCC spent more than 2 ns within 6 Å$<d_{drug-lipid}<$10 Å and away from lipids most of the time (see the upper panels in Figure 4). It suggests the possibility for drugs to briefly bind with lipids. Figure 5 (top panel) indicates that TXL likely favors the interaction with PC over PS while TCC shows no significant lipid specific preference. Both E_{ES} and E_{vdW} for TXL and TCC interacting with PC are inversely proportional to $d_{drug-lipid}$ while there are no such trends in either TXL-PS or TCC-PS cases. Below $d_{drug-lipid}<$6 Å which is on the order of the lipid head group dimension the CD-lipid binding stability drastically decreased (see the upper panel in Figure 5) and below $d_{drug-lipid}<$4 Å no stability was observed because there was no structure found at this low distance value (see Figure 2). It is shown that both E_{ES} and E_{vdW} are strongly effective within 12 Å ((E_{vdW} slightly dominant as shown in the inset plots of Figure 5).

3.1. Colchicine Kills Cancer Cells

Cell viability and cytotoxicity assays were used for drug screening and cytotoxicity tests of colchicine. Figure 6 indicates the colchicine's effect on cell viability. For initial screening, we chose breast cancer cells, for details see (Alqarni's M.Sc. thesis, supervised by Ashrafuzzaman, accepted by King Saud University, 2019). Methyl tetrazolium salt (MTT) assay checks viability based on various cell functions such as enzyme activity, cell membrane permeability, cell adherence, ATP production, co-enzyme production, and nucleotide uptake activity.

Many established methods such as Colony Formation method, Crystal Violet method, Tritium-Labeled Thymidine Uptake method, MTT, and WST methods are available. All these are based on counting the number of live cells. In MTT assay cell viability must be determined by counting the cells within a microscope after treatments. It is an Enzyme-based method using MTT rely on a reductive coloring reagent (formazan) and dehydrogenase (NADH) in a viable cell to determine cell viability with a colorimetric method.

Derivative drug stock solutions were prepared by weighing out a specific amount of the compound (colchicine, Col drug), dissolved in DMSO, and diluting to a final volume with DMEM media.

This method is far superior to the previously mentioned methods because it is easy-to-use, safe, has a high reproducibility, and is widely used in both cell viability and cytotoxicity tests. In this protocol, NADH reduces MTT to a purple formazan. The assay was based on the method described by Mosmann in 1983 [46].

Here we use the term 'the cells are less metabolically active due to drug effects' means the viability of cells is decreased and subsequently will reduce MTT color formation.

Figure 6 represents the result after treating breast cancer cells with concentration gradients of colchicine drug. Y-axis represents the cytotoxicity of MTT and X-axis represents the drug concentration starting from 10uM to 100uM. At 10uM colchicine, the cytotoxicity-started to increase means the drug (col) was toxic to the breast cancer cells, but non-significantly.

This toxicity was found to increase by increased concentration of drug until the cells became saturated at 100uM and the cytotoxicity decreased. Hansen and Bross [47] studied the putative toxic effects associated with the cultured cells exposure to a chemical substance including assays for cell viability and cell death. The viability of this study was used to examine the cytotoxic effects of celastrol. The cell-specific reduction of the MTT reagent was addressed by measuring the cellular viability; toxic effects was investigated by exposing of lymphoblastoid cells to the celastrol as a chemical drug and added MTT reagent to the cells and incubated the cells for 24 hour.

Dvorak et al. concluded that the biological activity of colchicine and its derivatives is tightly bound to their inhibitory effects on tubulin polymerization [48]. The major side effects of colchicine are demonstrated in proliferating cells, e.g., intestinal mucosa (nausea, vomiting) or hairs (alopecia). This implies that replacement of colchicine by the derivatives with lower anti-tubulin activity will result in lower side effects but also in lower therapeutic effects. However, this is only a hypothesis. It is shown that 10-*O*-demethylated derivatives of colchicine may display significantly lower cytotoxicity in human hepatocytes compared to colchicine. In addition, the tested derivatives may have distinct biological activity, which is primarily due to interference with microtubules.

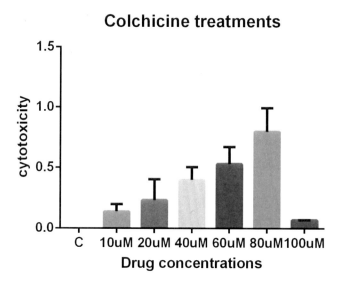

Figure 6. Variable concentrations of colchicine for cell viability detection. Cytotoxicity values were non-significant when compared to each other. The lower value for highest concentration tested here shows strange results. This is perhaps due to too high concentration of drugs might produce drug clustering, losing their effects on cell lines.

4. Discussion

4.1. The Evidence of Chemotherapy Drug Induced Ion Pore Formation in Lipid Membranes

The electrophysiology results indicate the formation of special structures inside membranes resulting from the action of the CD molecules

TCC and TXL. Erdal et al. have recently reported that a large number of potential anticancer drugs induce p53 protein-independent apoptosis and that lysosomal membrane permeabilization is a mediator of many such responses [49]. Our direct observation of both membrane permeabilization and the formation of ion pores due to TCC and TXL sheds new light on the cytotoxic effects and off-target interactions exhibited by these drugs. Stimulation of conductance events due to CDs may also lead to the discovery of a novel type of ion pore formed by compounds whose primary mode of action involves binding to cytoplasmic proteins. All membrane permeabilizing proteins or antimicrobial peptides form ion channels in which transitions between different current levels happen instantaneously. So far, only TCC and TXL have been shown to induce single conductance events with a current versus time plot that has a triangular shape. This indicates that the induced pore radius continuously changes back-and-forth giving a range of cross sectional areas of the pores. The time dependent change of the TCC/TXL pore conductance rules out the possibility of channel formation of the types represented by linear β-helix gA or barrel-stave Alm channels [20, 21]. Although the point count plots (see the right-hand panel of Figure S2.A) resemble random events observed for GS which represent defects [25], fine tuning of the conductance events (see Figure 1) suggests that they are long-lived, not just transient [25] but with no fixed conductance. No such time dependent transitions between non-zero current levels as observed in CD-induced discrete conductance events (see Figure 1) are found in GS-induced defects. The latter shows transient conductance events between zero current levels only. TCC/TXL events superficially resemble Alm events with important exceptions that there is an absence of stable and fixed Alm like discrete conductance values corresponding to different discrete channel cross sections representing different Alm conductance states. Also, unlike instantaneous transitions (apparently time independent) between Alm states, the CD induced pore current transitions are time dependent (see Figure 1). The CD induced pore conductance change (increasing and decreasing) over time was therefore found (presented in the Results section) to be finite while the peptide (gA or Alm) induced pore conductance changes over time appear with infinite values. Our results, however, can neither rule out nor support a mechanism similar to Bechinger's in-plane diffusion model [50, 51]. Here, the bilayer disrupting molecules disorder the hydrocarbon chains of the adjacent phospholipid molecules and create a local bilayer thinning. This leads to local disturbances in bilayer packing and an eventual formation of toroidal type pores (like those shown in Figure 7). It is worth mentioning that after bilayer formation with a combination of three different classes of lipids (phosphoethanolamine, phosphatydyleserine, phosphatidylcholine) we waited for 1 hr, tested the bilayer stability under at least 20 independent experimental conditions without the presence of drugs by applying a very high transmembrane potential ~ 400 mV for about 2-3 min. We found no instability or conductance events in the bilayer. Therefore, the creation of an electric field induced or any specific lipid induced conductance events can be ruled out which would be characterized as prepore or metastable states in lipid bilayers reported by e.g., Melikov et al. using a different lipid composition [52].

The observed conductance events were also determined not to be due to DMSO, as the maximum amount of it was less than 0.8% of the volume of the aqueous solution at which it does not induce any bilayer disrupting effects (confirmed in many of our studies and in ref. [53]). It is worth discussing whether these CD induced pores resemble the channels formed by ceramides in lipid membranes [54]. Ceramides, despite being lipids, form channels where the molecules are considered to be held together in a column spanning the hydrophobic portion of the membrane with the columns being held together along the channel lumen that makes a cylindrical pore (pore diameter ~ 0.8-11 nm) [54]. CD pore conductance changes spontaneously back-and-forth (see Figure 1) between lower and higher values. This indicates that no stable conductance levels exist as those observed in ceramide channel currents [54]. That means a number of ceramide molecule complexes created so-called 'lipidic' pores whose stable cross sections for a certain value of conductance resemble those in Alm channels. This rules out the possibility of any resemblance between CD induced pores and ceramide channels.

Colchicine Induced Ion Channel Formation into Membranes ... 173

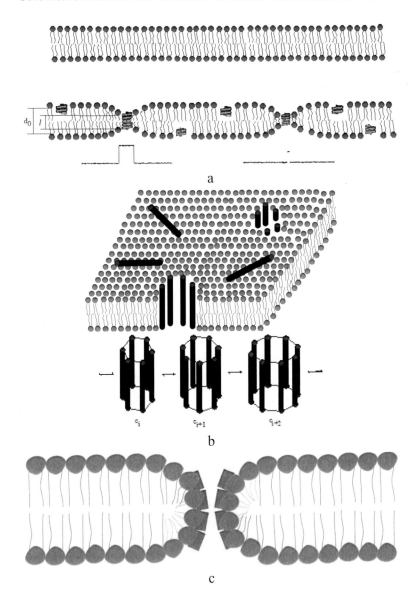

Figure 7. (A) gA channels (lower panel) deform lipid bilayer's resting thickness. d_0 and l represent hydrophobic bilayer thickness and gA channel length, respectively. (B) In Alm channels cylindrical rods represent monomers on a 'barrel stave' pore. Transitions between different Alm conductance pores by addition/release of monomer(s) from/to the surrounding space are shown in the lower panel. (C) Chemotherapy drugs TCC and TXL induced proposed toroidal-type ion pores in a lipid bilayer membrane possibly associated with a spontaneous time dependent change of the pore cross section.

TCC/TXL induced conductance events therefore do not appear to be similar to GS-induced defects or to other protein lined channels (e.g., gA's linear β-helix, Alm's barrel-stave pore etc., see Figure 7) or so-called 'lipidic' ceramide channels [54]. We have also ruled out the possibility of any observed conductance events to be due to membrane conditions such as those caused by an electric field across the membrane, any characteristic of membrane constituting lipids, DMSO, etc. The CD induced events therefore correspond to a special structure induced inside membranes by forming a drug/lipid molecular complex. We hypothesize that lipids are forced to line across pore openings (see Figure 7) and TCC/TXL molecules are localized behind the head group region near the hydrocarbon chains of the lipids [15]. The moderate dependence (linear) of A on c_D and V (see Figure S3) in comparison to the 2.6 power dependence on the identical parameters reported in the cases of protein-lined channels [55, 56] also provides support for the proposed model of the TCC/TXL pore (see Figure 7). This structure allows the membrane to be doped with higher amounts of molecules at a moderate energy cost as the molecules just penetrate into the lipid head group-hydrocarbon chain interface. Only this type of broken membrane structure can ensure a back-and-forth spontaneous change of the pore's geometric cross-sectional area requiring no addition or release of pore inducing agents which leads to the formation of triangular conductance events with a possibility of spontaneous change of pore conductance between any values. Admittedly, further studies are badly needed to confirm the proposed less defined stoichiometry and structure of the toroidal type channel which can possibly be induced by CDs in membranes. Work is currently underway in that direction using various experimental techniques and computer model studies. In the case of protein-lined channels the membrane's resting thickness near the channels is considered to be changed but may not vanish completely. This could indicate bilayer deformation as predicted in a mechano-sensitive channel study [57]. Unlike in gA channels [53] and Alm channels (submitted), which exhibit substantial phenomenological pH effects, we have not observed considerable pH effects on the drug induced pore activity (pH 5.7-8.5). This suggests that: (i) the phenomenological function of drug-induced pores is different from conventional protein-lined channels e.g., gA, Alm, etc. (see the model diagrams in Figure 7) and (ii) CD molecules are equally effective in both normal cell membranes and cancer cell membranes which exist in different pH environments [58]. Both of these points make sense considering that these molecules are electrostatic charge neutral while other channel forming peptides are charge bearing (pH sensitive) with specific (helical, β-sheet, etc.) structures.

4.2. The Drug-Lipid Physical Interaction as a Possible Cause of Chemotherapy Drug Induced Pore Formation

The main condition for any ion transporting pore/channel formation is that the channel forming agents and lipids must physically organize to form certain type(s) of stable structure(s). The purpose of performing MD simulations presented in this paper was to discover the physical energies that play primary role(s) to ensure the time dependent physical coexistence between CDs and lipids in membranes. MD simulations were meant not to calculate the stability of the toroidal pores rather as mentioned here, these simulations were designed to investigate the stability of the pairwise coexistence (due to physical interactions) of the channel forming agents and lipids. The stability of experimentally discovered toroidal pores is predicted to be due to resultant effects of many such CD-lipid physical interactions.

MD results suggest that the CD-lipid complex fluctuates within a separation over a period of time. These results suggest that both TXL and TCC likely bind with PC and PS given appropriate initial conditions. Through our in-depth analysis we found evidence suggesting that the hydrophobic effect is unlikely to contribute to the distance dependent CD-lipid binding. The analysis of energy contributions from two non-bonded interactions, E_{vdW} and E_{ES} versus $d_{drug-lipid}$ revealed crucial insights into the cause of the observed stability of the CD-lipid complexes. Both E_{vdW} and E_{ES} appear to be the main contributors to the energetic CD-lipid binding and vdW interactions contribute slightly more than ES interactions as the drug and lipid approach closer. Binding stability generally is found to decrease quickly with increasing $d_{drug-lipid}$ within 6 Å<$d_{drug-lipid}$<16 Å at which considerable structures were observed (see Figure 2). Both vdW and ES interactions contribute comparably with both energies decreasing with increasing $d_{drug-lipid}$. Beyond a 16 Å separation the interactions become negligible. The CD-lipid stability is still observed due to the effects of a combination of possible sustained presence of long range E_{ES} (though E_{vdW}≈0, data not shown) and other energies possibly induced by the surrounding environment. Large standard deviations (see Figure 5) are suggestive of the conformational space of the CD-lipid complexes not being completely explored in MD simulations. Nonetheless, this incompleteness does not preclude our proposed interpretation. Importantly, the CD-lipid interactions resemble the protein-lipid vdW and ES interactions found in MD simulations of a gA channel in phospholipid membranes [59]. This suggests that the CD-lipid interactions resemble these mechanisms, namely the vdW and ES interactions as found in the channel forming AMP-lipid interactions.

Both CDs are found to permeabilize lipid bilayer membranes by creating ion pores, a mechanism that is usually exerted on lipid membranes by various AMPs. The MD simulation based discovery of common types of primary physical interactions (vdW and ES) between pore inducing agents and lipids for both CDs and AMPs (will be published elsewhere) also suggest that these two different classes of biologically active molecules may act on membranes using common molecular actions. Therefore, CDs are likely to generate certain AMP type mechanisms.

A mechanistic understanding of the drug-phospholipid interaction can be gained from the novel picture of interaction energetics between membrane active agents (MAAs) and a lipid bilayer membrane that has recently been reported [29-32]. In this work, a screened Coulomb interaction model has been used to represent the interaction of MAAs with lipids in the bilayer, due to the presence of the distributions of the localized charges on both MAAs and phospholipids of all types. When any MAA and phospholipid approach each other to form a MAA-phospholipid complex (e.g., the case of CD-PS/PC interaction in this study) the localized charges present in the complex experience each other's electric fields. In this electrostatic interaction scenario the charges interact with each other not only directly (via Coulomb interactions) but also indirectly through other charges in the vicinity. Hence, the interaction between any two localized charges becomes screened, so the interaction takes the traditional form of the screened Coulomb interaction [60]. Furthermore, the interaction energy may change due to many other parameters such as variations in the membrane's electrical conditions, the presence of hydrocarbons, lipid density, etc.

In general, the binding energy between CD and phospholipid in a membrane is due to the sum of the traditional Lennard-Jones [32] and screened Coulomb potentials [29-32, 50]. A change of drug-phospholipid binding stability which appears through the distance-dependent probability (see Figure 5 in our MD simulation) is mainly due to the change of drug-phospholipid coupling energy which varies mainly according to the effects of the hydrophobic coupling to be filled with localized single, double etc. charges, respectively, that are present within the interaction field in a drug-phospholipid complex. In the case of the interaction of any CD with a phospholipid bilayer the screened Coulomb interaction extends beyond the nearest neighbor phospholipids to other phospholipids residing in the vicinity, exactly following the protocol presented in references [29-32, 60].

To better understand the generalized drug-PS/PC interactions, more MD studies are needed to focus on CDs and phospholipid membrane rather than a single phospholipid. This deeper understanding of the interactions will provide clear insights into the membrane effects of CDs.

4.3. Colchicine Shows Significant Cytotoxicity against Cancer Cell Lines

Cell viability and cytotoxicity assays were used for drug screening and cytotoxicity tests of colchicine. Figure 6 indicates the colchicine's effect on cell viability. Initial screening was performed on breast cancer cells. Colchicine effects on *in vitro* constructed lipid bilayer membranes or *in silico* colchicine-lipid binding energetics and kinetics offer crucial molecular level understanding of the colchicine effects on cellular systems. But actual cytotoxicity measurements presented here certainly offer the gross colchicine effects on cells in physiological environments. This study will certainly enhance our understanding of drug effects.

Conclusion

Chemotherapy drugs TCC and TXL permeabilize lipid bilayer membranes by forming ion pores. A possible cause of pore creation may be due to the resultant effects of many physical drug-lipid interactions that were discussed in our earlier publications. The reported results and our computational modeling contributed to an improved understanding of the general mechanisms of ion pore formation in membranes. These studies shed light on the underlying mechanisms of complex interactions between structurally stable (pore/channel forming) peptides or small biomolecules and liquid crystal structure type of lipid membranes. These study concluded that the molecular mechanisms of CDs include both the commonly investigated interactions in the cell's cytoplasm and the hitherto largely neglected interactions across the cell's membrane. The colchicine induced significant direct cancer cell line cytotoxicity assay results presented here add into our molecular mechanism claim. This new insight will be useful in finding novel drugs to treat cancer and other diseases by focusing both on the cytoplasmic and membrane regions.

Acknowledgment

We are thankful to G. Shaik for helping us get the cancer cell lines.

CONFLICT OF INTEREST

None

REFERENCES

[1] For a list of chemotherapy drugs see http://www.chemocare.com/bio/.

[2] Seidemann P, Fjellner B, Johannesson A. (1987) Psoriatic arthritis treated with oral colchicine. *J. Rheumatol.* 14: 777–79.

[3] Callen JP. (1985) Colchicine is effective in controlling chronic cutaneous vasculitis in lupus erythematosus. *J. Am. Acad. Dermatol.* 13: 193–200.

[4] Rosenman SJ, Ganji AA, Gallatin WM. (1991) Contact dependent redistribution of cell surface adhesion and activation molecules reorganization. *FASEB J.* 5: 1603.

[5] Mekory YA, Baram D, Goldberg A, Klajman A. (1989) Inhibition of delayed hypersensitivity in mice by colchicines: Mechanism of inhibition of contact sensibility in vivo. *Cell. Immunol.* 120: 330–40.

[6] Borisy GO, Taylor EW. (1967) The mechanism of action of colchicine: Colchicine binding to seaurchin eggs and the mitotic apparatus. *J. Cell. Biol.* 34: 533–48.

[7] Matsumoto G, Sakai H. (1979) Microtubules inside the Plasma Membrane of Squid Giant Axons and their Possible Physiological Function. *J. Membrane Biol.* 50: 1-14.

[8] Haga T, Kurokawa M. (1975) Microtubule formation from two components separated by gel filtration of a tubulin preparation. *Biochim. Biophys. Acta.* 392: 335.

[9] Agutter PS, Suckling KE. (1982) Effect of colchicine on mammalian liver nuclear envelope and on nucleo-cytoplasmic RNA transport. *Biochim. Biophys. Acta.* 698: 223-229.

[10] Holmes FA, Kudelka AP, Kavanagh JJ, Huber MH, Ajani JA, Valero V. (1995) Current Status of Clinical Trials with Paclitaxel and Docutaxel. In *Taxane Anticancer Agents: Basic Science and Current Status;* Georg, GI, Chen, TC, Ojima, I., Vyas, DM, Eds.; ACS Symposium Series No. 583; American Chemical Society: Washington, DC.; 31-57.

[11] Schiff PB, Fant J, Horwitz SB. (1979) Promotion of Microtubule Assembly *in vitro* by Taxol. *Nature.* 277: 665-66.

[12] Fisherman J, McCabe M, Hillig M. (1992) Phase I study of taxol and doxorubucin (Dox) with G-CSF in previously untreated metastatic breast cancer. *Proc. Am. SOC. Clin. Oncol.* 1175A.

[13] Sharma A., Straubinger, RM. (1994) Novel taxol formulations: Preparation and characterization of taxol-containing liposomes. *Pharm. Res.* 11: 889-96.

[14] Sonee M, Barron E, Yarber FA, Hamm-Alvarez SF. (1998) Taxol inhibits endosomal-lysosomal membrane trafficking at two distinct steps in CV-1 cells. *Am. J. Physiol. Cell Physiol.* 44: 1630-39.

[15] Balasubramanian SV, Straubinger RM. (1994) Taxol-lipid interactions: taxol-dependent effects on the physical properties of model membranes. Biochemistry. 33: 8941-47.

[16] Shiba M, Watanabe E, Sasakawa S, Ikeda Y. (1988) Effects of taxol and colchicines on platelet membrane properties. *Thromb Res.* 52: 313-23.

[17] Mons S, Veretout F, Carlier M, Erk I, Lepault J, Trudel E, Salesse C, Ducray P, Mioskowski C, Lebeau L. (2000) The interaction between lipid derivatives of colchicines and tubulin: Consequences of the interaction of the alkaloid with lipid membranes. *Biochim. Biophys. Acta.* 1468: 381-95.

[18] Jelokhani-Niaraki M, Hodges RS, Meissner JE, Hassenstein UE, Wheaton L. (2008) Interaction of Gramicidin S and its Aromatic Amino-Acid Analog with Phospholipid Membranes. Biophys J. 2008 October 1; 95(7): 3306–3321.

[19] Bond PJ, Khalid S. (2010) Antimicrobial and cell-penetrating peptides: structure, assembly and mechanisms of membrane lysis via

atomistic and coarse-grained molecular dynamics simulations. *Protein Pept Lett.* 2010 Nov;17(11):1313-27.

[20] Boheim G. (1974) Statistical analysis of alamethicin channels in black lipid membranes. *J. Mem. Biol.* 19: 277-303.

[21] He K, Ludtke SJ, Huang HW, Worcester DL. (1995) Antimicrobial peptide pores in membranes detected by neutron in-plane scattering. *Biochemistry.* 34: 15614-18.

[22] Andersen OS. (1983) ION MOVEMENT THROUGH GRAMICIDIN A CHANNELS Studies on the Diffusion-controlled association step. *Biophys. J.* 41: 147-65.

[23] Ashrafuzzaman M, Andersen OS. (2007) Lipid bilayer elasticity and intrinsic curvature as regulators of channel function: a single molecule study. *Biophys. J.* 421a.

[24] Ashrafuzzaman M, McElhaney RN, Andersen OS. (2008) One antimicrobial peptide (gramicidin S) can affect the function of another (gramicidin A or alamethicin) via effects on the phospholipid bilayer. *Biophys. J.* (2008) 94: 21.

[25] Ashrafuzzaman M, Andersen OS, McElhaney RN. (2008) The antimicrobial peptide gramicidin S permeabilizes phospholipid bilayer membranes without forming discrete ion channels. *Biochim. Biophys. Acta.* 1778: 2814-22.

[26] Matsuzaki K, Murase O, Tokuda H, Fujii N, Miyajima K. (1996) An antimicrobial peptide, magainin 2, induced rapid flip-flop of phospholipids coupled with pore formation and peptide translocation. *Biochemistry.* 35: 11361–68.

[27] Ludtke SJ, He K, Heller WT, Harroun TA, Yang L, Huang HW. (1996) Membrane pores induced by magainin. *Biochemistry.* 35: 13723–28.

[28] Yang L, Harroun T, Weiss TM, Ding L, Huang HW. (2001) Barrel-stave model or toroidal model? A case study on melittin pores. *Biophys. J.* 81: 1475-85.

[29] Ashrafuzzaman M, Tuszynski J (2012) Regulation of channel function due to coupling with a lipid bilayer. *J. Comput. Theor. Nanosci.* 9: 564-570.

[30] Ashrafuzzaman M, Tuszynski J (2011) Ion pore formation in lipid membranes due to complex interactions between lipids and channel formping peptides or biomolecules. In *HB of Nanosci., Eng. & Tech.*; Goddard, Brenner, Lyshevki and Iafrate, Eds.; Taylor and Francis (CRC press), New Yoek, USA; in press.

[31] Ashrafuzzaman M (2011) Antimicrobial peptides modulate bilayer barrier properties using a variety of mechanisms of actions, *Science against microbial pathogens: communicating current research and technological advances (Microbiology Book Series, No.3, Ed:* Antonio Méndez-Vilas*), Formatex Res. Cen., Spain*; Vol 2, 938-950.

[32] Ashrafuzzaman M, Tuszynski J (2012) Ion pore formation in lipid bilayers and related energetic considerations, *Curr. Med. Chem.* 19:1619-34.

[33] Min Yu, Aditya Bardia, Ben S. Wittner, Shannon L. Stott, Malgorzata E. Smas, David T. Ting, Steven J. Isakoff (2013). Circulating Breast Tumor Cells Exhibit Dynamic Changes in Epithelial and Mesenchymal Composition. *Science,* 339(6119):580-4.

[34] Edgardo Rivera and Henry Gomez (2010). Chemotherapy resistance in metastatic breast cancer: the evolving role of ixabepilone. *Breast Cancer Res.* 2010; 12(Suppl 2): S2.

[35] Chih-Yuan Tseng, Jonathan Y Mane, Philip Winter, Lorelei Johnson, Torin Huzil, Elzbieta Izbicka, Richard F Luduena3 and Jack A Tuszynsk (2010). Quantitative analysis of the effect of tubulin isotype expression on sensitivity of cancer cell lines to a set of novel colchicine derivatives. *Molecular Cancer,* 9:131.

[36] Lidia Śliwka, Katarzyna Wiktorska, Piotr Suchocki, Małgorzata Milczarek, Szymon Mielczarek, Katarzyna Lubelska, Tomasz Cierpiał, Piotr Łyżwa, Piotr Kiełbasiński, Anna Jaromin, Anna Flis, Zdzisław Chilmonczyk (2016). The Comparison of MTT and CVS Assays for the Assessment of Anticancer Agent Interactions. *PLoS One*, 11(5):e0155772.

[37] Sonia Arora, Xin I. Wang, Susan M. Keenan, Christina Andaya, Qiang Zhang, Youyi (2009). Inhibitor with potent anti-proliferative and anti-tumor activity. *Cancer Res.* 69(5): 1910–1915.

[38] Luciana Rizzieri, Luana Christine COMERLATO, Marcia Vignoli DA SILVA, José Îngelo Silveira ZUANAZZI, Gilsane Lino VON POSER, Ana Luiza ZIULKOSKI. Toxicity of Glandularia selloi (Spreng.) Tronc (2016). Leave extract by MTT and neutral red assays: influence o.f the test medium procedure. *Interdiscip Toxicol.* 9(1): 25–29.

[39] Ashrafuzzaman M, Chih-Yuan Tseng. 2018. Method for direct detection of lipid binding agents in membrane. Patent US9529006B1. https://patents.google.com/patent/US9529006B1/en.

[40] Case DA, Darden TA, Cheatham, TE III, Simmerling CL, Wang J et al. (2010) AMBER 11. University of California, San Francisco, USA.

[41] Wang J, Wang W, Kollman PA, Case DA (2006). Automatic atom type and bond type perception in molecular mechanical calculations. *J. of Mol. Graphics and Modelling.* 25, 247260.

[42] Wang J, Wolf RM; Caldwell JW, Kollman PA, Case DA. Development and testing of a general AMBER force field. *J. of Comp. Chem.* 2004;25: 1157-1174.

[43] Huzil JT, Mane J, Tuszynski JA. Computer assisted design of second generation colchicine derivatives, *Interdisciplinary Sciences-Computational Life Sciences.* 2010;2:169-174.

[44] Freedman H, Huzil JT, Luchko T, Luduena RF, Tuszynski JA. Identification and Characterization of an Intermediate Taxol Binding Site Within Microtubule Nanopores and a Mechanism for Tubulin Isotype Binding Selectivity, *J. of Chem. Info. and Modeling.* 2009:49;424-436.

[45] Ashrafuzzaman M, Duszyk M, Tuszynski J. (2011) Chemotherapy drug molecules thiocochicoside and taxol permeabilize lipid bilayer membranes by forming ion channels. *J. Phys.: Conf. Series.* 329, 012029, 1-16.

[46] Mosmann T (1983). Rapid colorimetric assay for cellular growth and survival: Application to proliferation and cytotoxicity assays. *J Imanol Meth.* 65:55–63.

[47] Hansen J, Bross P. A cellular viability assay to monitor drug toxicity. Methods Mol Biol. 2010;648:303-11. doi: 10.1007/978-1-60761-756-3_21.

[48] Dvorak Z, J Ulrichova, R Weyhenmeyer, V Simanek. 2007. Cytotoxicity of colchicine derivatives in primary cultures of human hepatocytes. *Biomed Pap Med FacUnivPalacky Olomouc Czech Repub.* 151(1):47–52.
[49] Erdal H, Berndtsson M, Castro J, Brunk U, Shoshan MC, Linder S. (2004) Induction of lysosomal membrane permeabilization by compounds that activate p53-independent apoptosis. *Proc Natl Acad Sci USA.* 102: 192-97.
[50] Bechinger B. (1997) Structure and functions of channel-forming peptides: magainins, sercopins, melittin and alamethicin. *J. Membr. Biol.* 156: 197-211.
[51] Bechinger B. (1999) The structure, dynamics and orientation of antimicrobial peptides in membranes by multidimensional solid-state NMR spectroscopy. *Biochim. Biophys. Acta.* 1462: 157-83.
[52] Melikov KC, Frolov VA, Shcherbakov A, Samsonov AV, Chizmadzhev YA, Chernomordik LV. (2001) Voltage-induced nanoconductive pre-pores and metastable single pores in unmodified planar lipid bilayer. *Biophys. J.* 80: 1829-36.
[53] Ashrafuzzaman M, Lampson MA, Greathouse DV, Koeppe II RE, Andersen OS. (2006) Manipulating lipid bilayer material properties using biologically active amphipathic molecules. *J. Phys.: Condens. Matt.* 18: S1235-55.
[54] Siskind LJ, Colombini M. (2000) The Lipids C2- and C16-Ceramide Form Large Stable Channels Implications for Apoptosis. *The J. of Biol. Chem.* 275: 38640-44.
[55] Boheim G, Kolb HA. (1978) Analysis of the multipore system of alamethicin in a lipid membrane. I. Voltage-jump current-relaxation measurements. *J. Memb. Biol.* 38 (1978)99-150.
[56] Latorre M, Alvarez O. (1981) Voltage-dependent channels in plannar lipid bilayer membranes. *Physiol. Rev.* 61: 77-150.
[57] Perozo E, Cortes DM, Sompornpisut P, Kloda A, Martinac B. (2002) Open channel structure of MscL and the gating mechanism of mechanosensitive channels. *Nature.* 418: 942-48.

[58] Gagliardi LJ. (2005) Electrostatic Considerations in Mitogenesis. *Proceedings ESA Annual Meeting.* 227-41.

[59] Woolf TB, Roux B. (1994) Molecular dynamics simulation of the gramicidin channel in a phospholipid bilayer. *Proc. Natl. Acad. Sci., USA.* 91: 11631-35.

[60] Ashrafuzzaman M., Beck H. (2004) in Vortex dynamics in two-dimensional Josephson junction arrays, (University of Neuchatel, http://doc.rero.ch/record/2894?ln=fr), ch 5, p 85.

APPENDIX: SUPPORTING INFORMATION, FIGURES

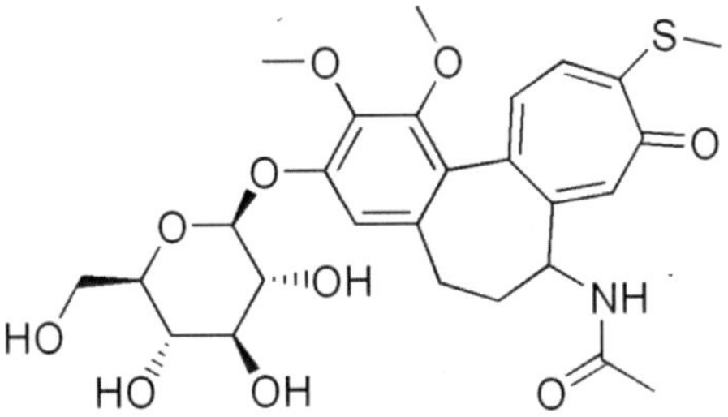

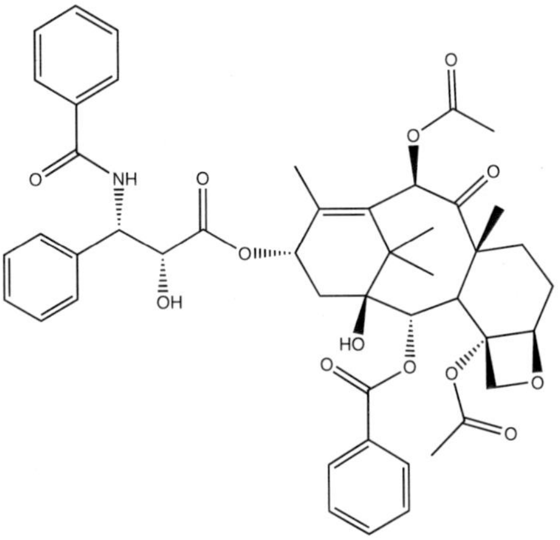

Figure S1. Structures of TCC and TXL (Taxol-A).

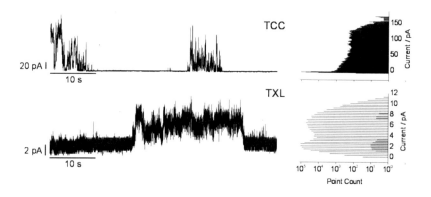

Figure S2.A. TCC and TXL (both at 90 μM) permeabilize lipid bilayer membranes by inducing nonzero current events. Here, pH=5.7, V=100 mV. Both traces were filtered at 20kHz but the lower one shows higher noise due to its presentation (current axis) at amplified scale.

Figure S2.B. A recorded 60 s current trace across a lipid bilayer membrane before being doped with any CD. No evidence was found on the presence of any conductance event like those observed in Figure S2.A. Here, pH=5.7, V=300 mV. The trace was filtered at 20kHz.

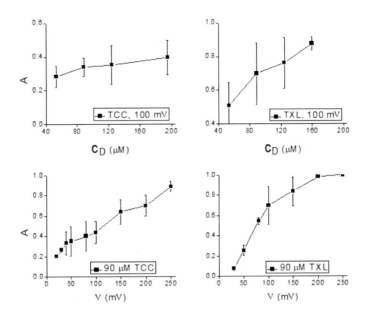

Figure S3. Activity A increases in proportion to V and c_D. $A=\sum A_i/(\sum A_i+A_{nc})$, A_i and A_{nc} are total point counts in the conducting phase i and the nonconducting phase, respectively. A_i and A_{nc} are therefore proportional to the times membrane behaves as a conductor and nonconductor respectively against membrane current. Each data point represents mean and standard deviation of the results recorded under three independent experimental conditions with three data recorded in each condition. pH=5.7.

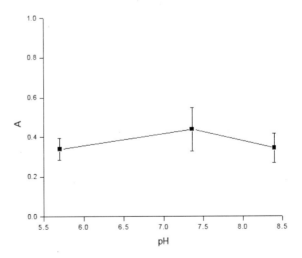

Figure S4. TCC pore activity A is independent of the pH of the membrane bathing aqueous phase. Here, c_D=90 µM, V=100 mV.

In: Lipid Bilayers ISBN: 978-1-53616-392-6
Editor: Mohammad Ashrafuzzaman © 2019 Nova Science Publishers, Inc.

Chapter 6

MOLECULAR STABILITY ANALYSIS OF REVERSE MICELLES ROLE IN BREAST CANCER DRUG DELIVERY AND IT'S DYNAMICS AND SIMULATION STUDIES

Dhivya Shanmugarajan[1,*], N. Premjanu[2], Ganesh Munuswamy[3], Lakshmi Jayasri Akkiraju[4], Sushil Kumar Middha[1] and Sureshkumar Chinanga[5,†]

[1]DBT-BIF Centre, Department of Biotechnology and Biochemistry, Maharani Lakshmi Ammanni College for Women, Bengaluru, Karnataka, India
[2]Sathyabama Dental College and Hospital, Chennai, Tamil Nadu, India
[3]Interdisciplinary Institute of Indian System of Medicine, SRM IST, Kattankulathur, Tamil Nadu, India
[4]Sri Padmavati Mahila Visvavidyalayam, Tirupati, India
[5]Barrix Agro Sciences Pvt. Ltd., Bengaluru, Karnataka, India

---

[*] Corresponding Author's Email: microroses@gmail.com.
[†] Corresponding Author's Email: csureshchem@gmail.com.

ABSTRACT

Statistics reveal breast cancer to be one of the foremost cancer affecting women worldwide. Almost 1 out of 4 women affected by cancer suffer from this form of cancer. Most of the anticancer drugs have poor solubility, limited permeability, reduced tumour targeting ability and decreased retention effect. Moreover, they produce detrimental effects in healthy tissues. To circumvent this problem, a promising approach is to selectively conjugate these drugs with polymeric reverse micelles as carriers for drug delivery. In this study, an *insilico* investigation on role of reverse micelles in drug-delivery for breast cancer has been carried out. The ligands /drugs are chosen and packed into the reverse micelles with different surfactants and studied for their stability, dynamics and simulation for 1 nanosecond. Further, screening of complex was carried out based on energy value and the best complex were subjected to the large-scale dynamics study to prove complex stability. Based on the result we are predicting that with similar environmental conditions of the drug with reverse micelles complex, they may be applied *invitro* after proper validation to improve chemo-drug pharmacokinetics and enhance aqueous solubility and bioavailability of the drug.

Keywords: reverse micelles, drug-delivery, *insilico*, surfactants, Dynamics and simulation

1. INTRODUCTION

1.1. Breast Cancer

Breast cancer is the main cause of mortality and the most common cancer in India accounting for about 27% of all cancers in women. The studies have indicated that the incidence of breast cancer within the age group of 50 to 64 years is more compared to other age groups, though the incidence of breast cancer starts in the early thirties itself. Moreover, the prevalence of breast cancer is higher in urban areas compared to rural areas. The ratio being 1:28 compared to 1:60 respectively [1].

The aetiology of breast cancer is multifactorial which includes either genetic or environmental or both. Some of the factors are mutation in

BRAC1 and BRAC2 gene, rapid urbanization, obesity, low basal metabolic index, waist to hip ratio, avoidance of breast feeding, improper diet, alcohol consumption, tobacco chewing, lack of exercise etc.

Female breast is composed of glandular tissue which contains glands for milk production connected to milk ducts through lobes and lobules. Apart from milk ducts breast contains adipose tissue, connective tissue and a well-developed network of lymph vessels, lymph nodes, blood vessels and nerves. Hence these tissues are highly sensitive to hormonal changes occurring during menstrual cycle, pregnancy or lactation. Breast cancer usually occurs from the inner lining of milk ducts or the lobules that provide the ducts with milk. They form a lump or mass of cells due to uncontrolled proliferation of cell called tumour. To locate these lumps for clinical examination the breast is divided into four sections inner upper section, outer upper section, inner lower section and outer lower section. Out of these sections the lumps are more prevalent in outer upper section of the breast near the arm pits [2].

1.2. Types of Breast Cancer

Breast cancer can be classified into three types:

1.2.1. Based on Tissues

1) Ductal carcinoma – in this type of cancer the inner lining cells of milk ducts are affected.
2) Lobular carcinoma - in this the cells involved in the production of milk i.e., lobules are affected.

1.2.2. Based on Invasiveness

1) Non-invasive carcinoma – It is also known as "in-Situ." The cancer does not undergo metastasis, hence does not spread to other parts of the breast.

2) Lobular carcinoma In-Situ (LCIS) - affects the lobules.
3) Ductal carcinoma In-Situ (DCIS) - affects the milk ducts.

1.2.3. Based on Hormones and Genes

1) ER - positive breast cancer - Estrogen receptor sensitive breast cancer
2) HER2 - Positive breast cancer - mutation in HER-2 gene, which is responsible for growth, division and repair of cells.

1.3. Challenges in Treatment of Cancer

Cancer being one of the major causes of mortality there is an urgent need to address this issue to curb the disease. The major obstacle is unawareness about breast cancer, general apathy towards female health, scarcity of diagnostic aids and to a certain extent financial constraint. In general cancer treatment involves radiation, surgery and chemotherapy. The chemotherapy being the most common and prevalent one. But chemotherapy has its own limitations like poor solubility, toxicity, restricted permeability, reduced bioavailability, low tumour targeting, inadequate cellular drug uptake, low curative efficacy, reduced retention effect, and cytotoxicity to healthy tissues. To circumvent this problem, a promising approach is to selectively conjugate these drugs with polymeric reverse micelles as carriers for drug delivery.

1.4. Lipid-Based Drug Delivery Systems

Lipids can be used as a suitable candidate for creating micelles for drug delivery. Lipids have various advantages such as they are bio-compatible, versatile, safe, efficient, stable and cost effective. Moreover, they are capable of site-specific targeting as well as time-controlled delivery of drugs to the target cell. These Lipid-based preparations can be customized according to the requirements of the diseased tissue and route of administration whether it is oral, parenteral, dermal/transdermal, or vaginal. For the lipid to be utilized as a vehicle for delivering the drug it should satisfy certain conditions such as

1) Solubility
2) Miscibility
3) Stability at room temperature
4) Solvent efficiency
5) Digestibility
6) Compatibility
7) Ability to aid self-dispersion
8) Absorbability
9) Economicalin preparation

For lipid drug delivery vehicle, it has to be attached to one or more surfactants such that lipids form the main ingredient and these surfactants helps in miscibility, solubilisation, and dispersion. Surfactants are classified based on their ability to balance hydrophobic and hydrophilic nature of the molecule. It is termed as hydrophilic - lipophilic balance number (HLB). If HLB number value is less than or equal to 10 then it is considered as lipophilic and if the value is more than 10 it is considered hydrophilic. Based on this various type of lipid drug delivery system can be formulated [3]. Lipid based drug delivery system can be classified as follows.

1.4.1. Emulsions

Emulsions are mixture of two immiscible liquids stabilized by an emulsifying agent. The two liquids present will be in such a way that one will be in continuous phase and another in discontinuous phase or so-called dispersed phase. The characteristic properties of emulsion depend on properties of dispersed phase and the continuous phase. In the emulsion there is interface which is a border line between dispersed and continuous phase. These emulsions can be further divided into Microemulsion, Self-emulsifying drug delivery system (seeds), Nanoemulsion, Pickering emulsion.

1.4.2. Microemulsion

Microemulsions are nano-sized isotropic system made up of aqueous and oily phase in the presence of large amount of surfactant in a proper proportion. The microemulsion size ranges below 200nm and it has low interfacial tension. Microemulsions have low viscosity, but high drug solubility, surface activity, flux rate, and enhancement ratio. It is also having mechanical and thermal stability. Moreover, they are formed spontaneously because of these properties it forms a suitable candidate to deliver drugs o the target cells [4].

1.4.3. Self-Emulsifying Drug Delivery System (Seeds)

These are self - emulsifying isotropic mixture that occurs when the entropy changes are greater in favour of dispersion rather than to increase surface area of dispersion. The emulsifying process is dependent on the concentration of surfactant, nature of the oil- surfactant pair, and the temperature at which it occurs. It is mainly used to enhance the oral absorption of highly lipophilic compounds. The main advantage of seeds in drug delivery is requirement of less energy for emulsion formation. Easy manufacturing process, Versatile and high drug payloads ranging from less than 25 mg to greater than 2000mg, shielding of sensitive drugs, delivering the drug directly to the oxygenated blood and enhancing the bioavailability of lipophilic drugs by increasing the gastric retention time [5].

1.4.4. Nanoemulsion

Nanoemulsions are colloidal particulate system made up of oil, emulsifying agents, and aqueous phase. Choosing of surfactant is the most important part in the production of nanoemulsion. The concentration of surfactant must be high, also it should have ultra low interfacial tension as well as flexible to promote and stabilize nanoemulsion. Their size ranges from 10 to 1000nm. They are solid spheres with amorphous surface and lipophilic with negative charge. Three type of nanoemulsion can be produced:

1) Oil in water nanoemulsion
2) Water in oil nanoemulsion
3) Bi-continuous nanoemulsions.

The advantage of using nanoemulsion in drug delivery is it improves bioavailability, enhances solubilisation of lipophilic drug, non-toxic, increased physical stability, has greater surface area providing greater absorbtion, can be formulated in different forms such as foams, creams, liquids and sprays. Moreover, can be used for targeted drug delivery of different anticancer drugs and also provides prolonged action of the medicaments [6].

1.4.5. Pickering Emulsion

Pickering emulsion is an emulsion of two immiscible liquids stabilized by solid particles at the interface. Depending upon the wettability of solid particles at the interface the continuous phase and dispersed phase is determined. The liquid which wets more is considered as continuous phase and the other as dispersed phase. These emulsions can interact with loaded drugs which in turn can release the drug at specific sites in a sustained manner [7].

1.5. Vesicular System

The vesicular carriers are exceptionally ordered association of bilayers of lipid. When amphipathic molecules of bilayer lipids encounter water, they form single or concentric spheres. The speciality of vesicular system is because of their amphipathic nature it has the capacity to hold both hydrophilic as well as hydrophobic drug. Also efficiently reduces drug toxicity by targeting to the specific site. Moreover, due to vesicular nature it protects the drug from degradation, oxidation etc. and aids in sustained release of drug. The nature of vesicular carrier depends on the concentration, composition, configuration, entrapment, shape and charge of the molecule. Based on the composition different type of vesicular systems are there such as Liposomes, Niosomes, Pharmacosomes, Phytosomes, Ethosomes, Transferosomes, and Vesosomes [8].

1.6. Lipid Particulate System

1.6.1. Lipospheres

Lipospheres are lipid-based water dispersible solid core encompassed by a single unit of phospholipid layer. The size of the solid particle ranges between 0.01 and 100 µm in diameter. The solid lipid hydrophobic core contains drug molecule which is covered by phospholipid layer which gives stability. Advantage of using lipospheres as drug delivery system is enhanced stability of the drug in the formulation, high-drug load, reconstitution properties, controlled particle size and drug release as well as its non-toxic nature. Aditionally, lipospheres protect the drug candidates from degradation, hydrolysis, increase the shelf life enabling high bioavailability and sustained plasma levels [9].

1.6.2. Lipid Drug Conjugates

They are drug molecules which have been modified covalently with the lipids. This association of lipids to drug molecules enhances lipophilicity and changes other properties of drugs. The conjugates exhibit several advantages which include improved oral bioavailability, improved targeting to the lymphatic system, enhanced tumour targeting, and diminished toxicity. Based on the chemical nature of drugs and lipids, different combination strategies and chemical linkers can be utilized to synthesize LDCs. The release of drugs depends on conjugation methods and linkers used which is very crucial for the ideal performance of LDCs [10].

1.7. Reverse Micelles Role in Drug Delivery

Reverse micelles are lipid molecules that arrange themselves in a spherical form in organic solutions Reverse micelles are nanometer-size droplets of aqueous phase, stabilized by surfactants in an organic phase. Reverse micellar systems have been developed using various organic and aqueous phases and surfactants. In reverse micelle hydrophilic groups make up the core and hydrophobic groups are present on the surface layer. The formation of a micelle is a response to the amphipathic nature of fatty acids, Reverse micellesis a thermodynamically stable system; it forms spontaneously and is fully transparent. Additional utilization of RM lies in their ability to act as hydrophilic nano-reservoirs dispersed in a hydrophobic phase. They are biocompatible, versatile, safe, efficient, stable and cost effective. Moreover, they are capable of site-specific targeting as well as time-controlled delivery of drugs to the target cell [11].

1.8. Dynamics and Simulation

Molecular dynamics (MD) is a form of computer simulation in which atoms and molecules can interact for a period. Because molecular systems generally consist of a vast number of particles, it is impossible to find the properties of such complex systems analytically; MD simulation circumvents this problem by using numerical methods. It represents an interface between laboratory experiments and theory. It calculates the time dependent behaviour of molecular system. Provides detailed information on the fluctuations and conformational changes of proteins and nucleic acids. It is used to investigate the structure, dynamics and thermodynamics of biological molecules and their complexes.

In molecular dynamics force fields are used. Force field refers to the functional form and parameter sets used to describe the potential energy of a system of particles. A Force- Field is assigned to each atom in the protein. The four key contributions to a molecular mechanics force field: are bond Stretching energy, bending energy, Torsion Energy and Non-bonded

Energy. So, the total energy of atom depends on = Stretching Energy + Bending Energy + Torsion Energy + Non-Bonded Interaction Energy. From which through statistical mechanics thermodynamic properties can be calculated. By this computational simulations questions relevant to drug delivery system design can be answered. Computer modeling can be utilised to examine the driving forces for nanoparticle membrane interactions, and how various parameters affect it. These parameters include size, shape, surface chemistry, and concentration, distribution, rigidity of nanoparticles, as well as mechanical and elastic properties of both nano particles and membranes and their affinity to receptors can be studied.

Computer simulations can be used to study self-assembly, the structural and dynamical characteristics of the resulting aggregates, drug loading capacity, drug distribution and localization in drug delivery system, complex stability, drug retention, release mechanism and release rate, dominant drug–drug delivery system interactions and to design or optimize DDSs targeting capabilities environmental conditions, e.g., pH, temperature, salt type and concentration, counter ions and external stimulus such as external magnetic fields, as well as interactions with other biomolecules [12].

2. MATERIALS AND METHODS

2.1. Packing of Lipid with Drugs

Drug delivery process involves insertion or packing of the drug in drug carrying vehicles. Dipalmitoylphosphatidylcholine (DOPC) lipid system of molecular weight 734.03 g/mol is widely used as a drug carrier. In the current *insilico* investigation DOPC was converted into a form of reverse micelles [13]. The converted form of DOPC was minimized and processed for 4ns stability process using CHARMm force field [14]. The optimized structure of DOPC was used for packing drugs used in the treatment of breast cancer. Drugs like Palbociclib, Doxorubicinm Gemcitabine, Abemaciclib were packed into DOPC using tool PACKMOL (http://m3g.iqm.unicamp.

br/packmol/) [15]. Major application of PACKMOL is to pack molecules in defined space. The tool also helps prevent short range repulsive interactions with other molecules during molecular dynamics simulations [16]. An illustrate code of packing reverse micelle with drug of interest is stated as follows

```
tolerance 2.0

structure Doxorubicin.pdb
    number 1
    fixed 0. 0. 0. 0. 0. 0.
    centerofmass
end structure

structure DOPC.pdb
    number 10
    atoms 41
        inside sphere 0. 0. 0. 10.
    end atoms
    atoms 99
        outside sphere 0. 0. 0. 10.
    end atoms
end structure

structure DOPC.pdb
    number 7
    atoms 41
        inside sphere 0. 0. 0. 10.
    end atoms
    atoms 99
        outside sphere 0. 0. 0. 10.
    end atoms
end structure

output Sample.pdb
```

2.2. Energy Minimization

Packed molecules are subjected to two stage of minimization process with steepest descent (SD) and conjugate gradient (CG). SD is first derivative and a slow convergence method of saving the position of molecule coordinate to local minima [17]. In other words, it converges slowly to minimum potential energy from the complex energy surface. Let the consider the following equation

$X(fp) = X(p1), X(p2), X(p3) .... X(pn)$

where X is the molecule, fp is final position of the coordinate, p1, p2, p3, ..., pn are the 1st, 2nd, 3rd and nth of the position coordinates. The position of molecule projects slowly forms p1 to p2 and then p3 till it reaches pn where it generates the final potential energy of the molecule (Figure 1A).

In the next process SD output is fed as an input for the second step of minimization in CG. CG is faster than SD and mostly used for fast convergence of molecule, but it requires the history of previous minimization steps [18]. It follows the simple equation as stated as below.

$$X(fp) = X(p1), ... X(p20), X(p23), X(p25), X(p27) .... X(pn)$$

where X is the molecule, fp is final position of the coordinate, 1st, 20th, 23rd and nth are the position coordinates. Position of molecule convergence is faster from p1 to p20 and then slowly moves from p21 to p23 finally reaching pn and generates the final potential energy of the molecule (Figure 1B). As like SD, CG is also an iterative method. Accordingly, the drugs were packed with reverse micelles aggregates of 15, 17 and 22 then processed individually following the two stage of minimization process prior to molecular dynamics simulation.

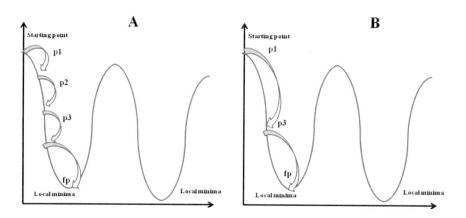

Figure 1. Energy minimization process of steepest descent (SD) and conjugate gradient (CG).

2.3. Molecular Dynamics Simulation

Molecular dynamics was processed in two phases, in the first phase each drug was packed with different aggregates of reverse micelles such as Palbociclib+RM15, Palbociclib+RM17, Palbociclib+RM22, Doxorubicinm+RM15, Doxorubicinm+RM17, Doxorubicinm+RM22, Gemcitabine+RM15, Gemcitabine+RM17, Gemcitabine+RM22, Abemaciclib+RM15, Abemaciclib+RM17, Abemaciclib+RM22. Then the drug + reverse micelles were subjected for 1000ps or 1ns short scale dynamics process in three steps.

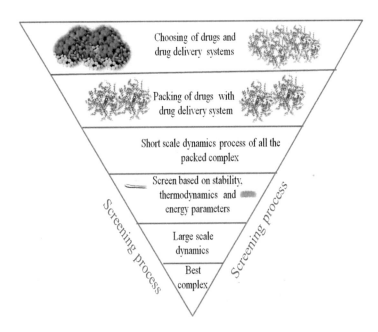

Figure 2. Virtual screening of drugs and vehicle in drug delivery process.

The first step involves raising the temperature of complex systems from 50k to 300k for 4ps simulation time, this is followed by equilibration for 500ps, during this equilibration time all the atoms will try to attain their degree of freedom. Finally, production was carried out using NVT ensemble with target temperature of 300k, electrostatics interaction of spherical cutoff was introduced along with thermal mass in the Hoover constant temperature

(Tmass) and Langevin piston for constant pressure control (Pmass) of 1000 with collision frequency of the Langevin piston was optioned as 25. Throughout dynamics SHAKE constraint algorithm was used to fix all bonds and angles with leapfrog dynamics integrator. Consequently, the more stable complex was processed for 15ns dynamics process to confirm its stability with respect to time. The stability of complex was studied using energy parameters such as electrostatic energy, potential energy and RMSD (Root mean square deviation). An overview of the screening process in insilico approach towards drug delivery is illustrated in the Figure 2.

3. RESULTS AND DISCUSSION

3.1. Significance of Minimization

Minimization is a very crucial process to optimize the structure of proteins, lipids, nucleic acids and chemical structures before dynamics simulation. During energy minimization each atom will try to attain the least energy and best conformation as possible in the molecule. Based on the system implication force field, energy minimization will vary, for instance, chemical structures optimization requires robust short scale minimization step as compared to macromolecules. Moreover, lipid system needs more attention on minimization technique when packed with proteins or chemical structures.

Dhivya et al., earlier described detailed packing of lipid bilayer with GPCR proteins using perl script, after each step of scaling lipid bilayer with protein, energy minimization was carried out to avoid the bad clashes and steric contact between the atoms [19]. Similarly, a study by Mackay explained the role of energy minimization for bimolecular systems and explains about the minimization Strategies before simulation [20]. The current study involving breast cancer drugs, the drugs were packed with reverse micelles followed by energy minimization to prevent internal clashes of atoms.

However, it was observed during packing with packmol tool that some of the drugs showed internal bad clashes with reverse micelles, in those cases the volume of drug placing coordinates were optimized and then packed with reverse micelles.

Table 1. Potential energy of Drug with reverse micelles aggregate during minimization process

Drugs	Reverse micelle Aggregate	Minimization	Initial Potential Energy (kcal/mol)	Final Potential Energy (kcal/mol)
Palbociclib	15	SD	17885.894	-1012.132
		CG	-1012.132	-2653.684
	17	SD	17702.946	-731.548
		CG	-731.548	-2591.59
	22	SD	551158.425	-774.601
		CG	-774.601	-3537.016
Doxorubicin	15	SD	1540717433	-912.924
		CG	-912.924	-2373.305
	17	SD	1530719083	-985.835
		CG	-985.835	-2784.251
	22	SD	1560785871	-1485.024
		CG	-1485.024	-3927.409
Gemcitabine	15	SD	729225.285	-872.728
		CG	-872.728	-2689.63
	17	SD	878181.943	-975.589
		CG	-975.589	-3081.325
	22	SD	957641.36	-1489.576
		CG	-1489.576	-4356.527
Abemaciclib	15	SD	96362.776	-659.092
		CG	-659.092	-2297.281
	17	SD	99074.86	-894.574
		CG	-894.574	-2687.712
	22	SD	2069607.636	-1267.118
		CG	-1267.118	-3953.628

Initial energy of the drug molecule packed reverse micelles was calculated for all the complexes. All the drug complexes showed positive unstable potential energy (Table 1). The dual process of minimization with SD followed by minimization with CG showed much more favorable structural stability of the complex compared to the minimization with SD alone. Since, structures obtained by first stage minimization using SD showed considerably negative potential energy. In the current study it was observed that drugs packed with maximum aggregate RM22 were more stable than reverse micelles aggregates RM15 and RM17 (Table 1). Moreover, Gemcitabine+RM22 complex had the most stable potential energy value of -4356.527 kcal/mol, followed by Abemaciclib+RM22 (-3953.628 kcal/mol) and Doxorubicin+RM22 (-3927.409 kcal/mol). There is a possibility that increasing the number of aggregates might yield a more stable complex, but that is beyond the scope of our current study, which is limited to three trial aggregate numbers.

3.2. Thermal Stability of the Complexes

Thermodynamic stability of the complex is more crucial in drug delivery, the drug with reverse micelles complex were subjected for heating process as a first step of simulation. The reverse micelles complexes were subjected to increasing temperatures, starting from 50k and ending at 300k. During the heating phase, the temperature was gradually increased by assigning increasing random velocities to each atom at predetermined time intervals, at the end of heating phase each complex employed in the study will try to attain its maximum stable temperature. The change in temperature is dependent on time of dynamics, in each interval time difference in temperature was noted, but the variations were not significantly deviated for any of the drug with reverse micelles complex.

Molecular Stability Analysis of Reverse Micelles Role ...   205

However, some complex maintained their temperature stability. Figure 3 illustrates temperature profile of Palbociclib, Doxorubicin, Gemcitabine and Abemaciclib reverse micelles complex. It is clear from the results that the drugs packed with RM22 have not deviated above 315k whereas, other complexes of drugs with RM15 and RM17 have undergone temperature fluctuations with respective to time.

Consequently, it was observed that breast cancer drugs with reverse micelles RM22 are more favorable for further *invitro* analysis. Temperature is considering an important parameter for drug delivery, increased deviation in varying temperatures will have an appreciable effect on the *invitro* results, thus only aggregates with stable temperature can be used for experimental analysis.

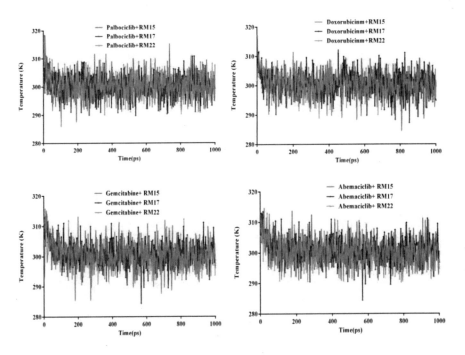

Figure 3. Temperature profile of reverse micelles with drugs.

3.3. Energy Stability

During dynamics various energy parameters will decide the stability of the complex which includes potential energy, kinetic energy, electrostatic energy and vanderwaals energy. Each energy has its own role in making a stable complex. Michael Levitt calculated the thermodynamic and structural properties of molecules by performing molecular dynamics simulation using an analytical expression for potential energy of system [21]. Potential energy of the system was calculated based on position taken by each conformation. Basically, the function is a simple empirically derived mathematical expression that gives energy of the complex as a function of atoms, also the energy of the system will vary from one point to another point until it attains a stable conformation.

Moreover, molecular mechanics is based on empirical potential energy functions, which is widely used to study macromolecules such as proteins, lipids, peptides and chemical structures. We investigated the potential energy of drugs packed with different aggregates of reverse micelles (RM). The energy of Doxorubicin+ RM and Gemcitabine+RM follows the same pattern as compared with Abemaciclib+RM and Palbociclib+RM (Figure 4). The drug +RM complexe's Palbociclib+RM15, Palbociclib+RM17, Abemaciclib+RM15 and Abemaciclib+RM17 showed potential energy values in the range of ~ -1000kcal/mol to ~ -2000kcal/mol. However, for the complexes Palbociclib+RM22 and Abemaciclib+RM 17 potential energy values starts form ~ -2000 kcal/mol and attains the stability before reaching the -3000kcal/mol. Whereas, Gemcitabine+RM22 energy values starts from -2500 kcal/mol and follows a stable decreasing pattern until -3200 kcal/mol. Thus, among all of the chosen potential energy complexes, Gemcitabine+RM22 and Doxorubicin+RM22 are the most stable.

Kinetic energy is also one of the most common energy parameters in molecular dynamics simulation, it is calculated from the atomic velocities. The kinetic energy value of RM15 with drugs Palbociclib, Gemcitabine, Doxorubicin, and Abemaciclib starts from 1550 kcal/mol are lower compared to RM17 complexes whose values begins at 1800 kca/mol. However, RM22 drug complexes have their kinetic energy values in the range of 2200 to 2500kcal. It was observed that the system does not show maximum deviations and each complex follows the stable equilibrium.

Electrostatic interactions are non-bonded interactions computed using the partial charges assigned to the atoms in the system. This energy is highly contributed to study conformational stability, biomolecular folding and binding energies of biomolecules. Figure 5 illustrates the electrostatic energy of the drug reverse micelles complex. The drugs Doxorubicin and Gemcitabine packed with RM22 have their energy values are in the range of -5,000 kcal to -5500 kcal/mol. Whereas, Palbociclib and Abemaciclib packed with RM22 have values in between -4500 kcal/mol and -5000 kcal/mol. These values are less favorable than the energy values of Doxorubicin and Gemcitabine packed RM22 complexes. However, the least favored complex systems were RM15 and RM17 packed drug complexes within the energy value of -3000 kcal/mol to -4,00kcal/mol.

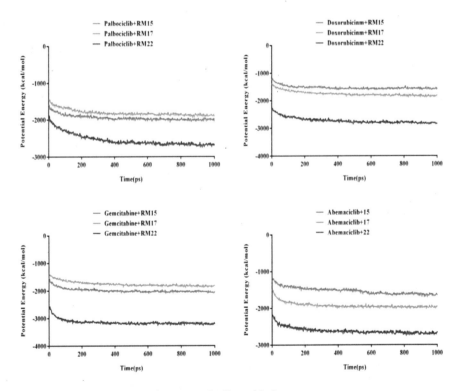

Figure 4. Potential energy of reverse micelles with drugs.

A non-covalent bond interaction during simulation was calculated as vanderwaals energy. This energy is approximated in the CHARMm empirical energy function by the Lennard-Jones potential energy function. This function is often referred to as a 6-12 potential because the attractive force is proportional to R-6, while the repulsive force is modeled by R-12. Vanderwaals interaction of drug complex Palbociclib (RM15, RM17) and Gemcitabin (RM15, RM17) showed similar energy value curves that closely match each other values. On the other hand, the RM22 complex drugs had energy values starting at -1000 kcal/mol and attain stability throughout the dynamic's simulation. Hence, by considering all the energy parameters, Gemcitabine+RM22 and Doxorubicin+RM22 were considered to be more energetically stable complex, thus these two complexes were taken for further large-scale molecular stability studies using CHARMm force field.

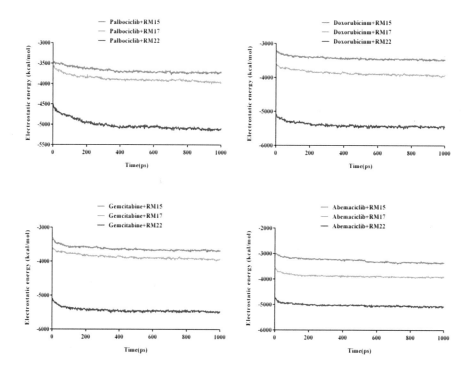

Figure 5. Electrostatic energy of reverse micelles with drugs.

3.4. Large Scale Dynamics of the Best Complex

Among twelve different complexes, only the best stable energy complex in short nanoscale dynamics was taken for further 15ns molecular dynamics simulation. Structural variation in the drug delivery complexes Doxorubicinm+RM22 (Figure 6A) and Gemcitabine+RM22 (Figure 6B) during 0ns, 5ns, 10ns and 15ns was observed (Figure 6).

The fluctuation equilibrium and structural stability of the Doxorubicin+RM22 and Gemcitabine+RM22 during Molecular dynamics (MD) simulations were examined by monitoring the backbone root mean square standard deviation (RMSD) with respect to their starting structures as a function of simulation time. Longer trajectory showed that these complexes stay stable and does not show other abrupt increases or decreases in their RMSD value. RMSD curves of Doxorubicin+RM22 (Orange line) and Gemcitabine+RM22 (black lines) showed a steady increase during the first 3000ps (Figure 7A).

Afterwards, these two curves exhibited a relatively stable fluctuation with the fluctuation amplitude of less than ~0.4Å from 5000ps to 10000ps thus indicating that both curves have reached equilibrium. Further, with increase in time, MD simulations of Doxorubicin+RM22 and Gemcitabine+RM22 got merged at ~14500ps and at the end of 15000ps Doxorubicin+RM22 RMSD values were slightly lower compared with Gemcitabine+RM22. Overall, both the systems are stable as per the structure analysis in terms of time.

In the real time studies temperature is crucial parameter, theoretical prediction will seed the knowledge about the stability of complex with respect to thermal equilibrium. Also, through this type of study randomness in the process is prevented, instead it specifically reveals about the implementation of accurate temperature, thereby, avoiding experimental uncertainty.

Temperature deviations in complexes Doxorubicinm+RM22 and Gemcitabine+RM22 was not observed above the upper limit of ~310K and lower limit of ~290K, hence, room temperature is considered to be more favorable for these complexes during drug delivery, however, a slight decrease in temperature also seems to favor the systems (Figure 7B).

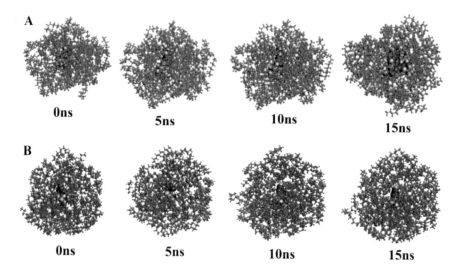

Figure 6. A: Doxorubicinm+RM22 structural variation based on time during MD simulation, B: Gemcitabine+RM22 structural variation based on time during MD simulation.

In regard to energy parameters, potential and electrostatics interactions are the top priority because both these energies are related with biological process. Potential energy values of Gemcitabine+RM22 begins with stable value of -2500 kcal/mol and it steadily decreased at 3000ps, further expansion of simulation process follows the same pattern without any distortion in the energy values. The trajectory path of Doxorubicin+RM22 starts with a least value of -2200kcal/mol as compared with Gemcitabine+RM22, but a slight decrease in curve was observed at 3500ps (Figure 7C) which tends to attain stable energy curve values at the end of dynamics.

Comparatively Gemcitabine+RM22 complex has a steady and more stable energy pattern than Doxorubicin+RM22. Electrostatic interaction curve of Gemcitabine+RM22 and Doxorubicin+RM22 follows the same trail of ~1000kcal energy variation (Figure 7D) with values of Doxorubicinm+RM22 showing slight elevation at the beginning compare to the Gemcitabine+RM22 complex values, but both curves achieved the stability without any abrupt changes in energy patterns.

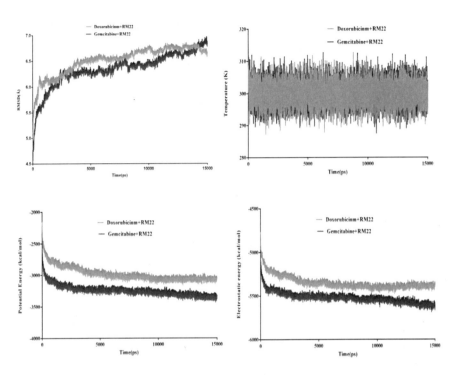

Figure 7. Large scale dynamics and Simulation A: RMSD, B: Temperature profile, C: Potential energy, D: Electrostatic energy.

CONCLUSION

Targeted drug delivery to the specific site of action is a very important issue in medical and pharmaceutical research. In this aspect, computational biology tools are boon to modern research. They help researches use simulation-based studies to identify potential drug delivery systems suitable for delivering drugs to the target site. This helps avoid randomness in the experimental process and provides researchers with a narrowed down starting point, thus saving researches time and other valuable resources. Computational modeling strategy described in this chapter reveals important aspects of the reverse micelles with anti-cancer drugs, a valuable tool for cancer therapy. From the current study, we can gain insight into the DOPC based drug delivery system. The study reveals that not all the drugs packed with reverse micelles are stable in drug delivery. The current chapter deals with a typical complex behavior and details about small- and large-scale molecular dynamics simulation. The chapter also shows the estimation of temperature for the complexes and use of this method in context to the experimental design. Though the present study was limited only with four drugs and three different aggregates of reverse micelles the data shows that the study can be used for any drugs packed with different types of drug vehicle packing material such as micelles, reverse micelles, liposome and other drug delivery vehicles. The pattern and results will change from one system to another, but knowledge gained by this study could definitely be used to improve the design of drug delivery system and it can save ample time in biological and pharmaceuticals research.

REFERENCES

[1] Chaurasia, V. & Pal, S. (2014). A Novel Approach for Breast Cancer Detection using Data Mining Techniques. *International Journal of Innovative Research in Computer and Communication Engineering, 2(1)*, 2456-65.

[2] Muhammad, A. M. I, Muhammad, D, Asmat, U. K. (2017). Awareness and current knowledge of breast cancer. *Biol. Res. 50:33*.

[3] Hina, S, Rajni, B, Sandeep, A. (2014). Lipid-Based Drug Delivery Systems. Journal of Pharmaceutics, Article ID 801820, 10 pages.

[4] Anoop, K, Vibha, K, Pankaj, K. S. (2014). Pharmaceutical Microemulsion: Formulation, Characterization and Drug deliveries across skin. *Int. J. Drug Dev. & Res., 6(1)*, 1-21.

[5] Gajendra, S, Khinchi, M. P, Gupta, M. K., Dilip, A, Adil, H, Natasha, S. (2012). Self emulsifying drug delivery systems (SEEDS): An approach for delivery of poorly water soluble drug. *Int. J. of Pharm. & Life Sci., 3(9)*, 1991-1996.

[6] Manjit, J, Rupesh, D, Sharma, P. K. (2015). Nanoemulsion: an advanced mode of drug delivery system. *3 Biotech*, *5(2)*, 123–127.

[7] Yunqi, Y, Zhiwei, F, Xuan, C, Weiwang, Z, Yangmei, X, Yinghui, C, Zhenguo, L, Weien, Y. (2017). An Overview of Pickering Emulsions: Solid-Particle Materials, Classification, Morphology, and Applications. *FrontPharmacol*, *8*, 287.

[8] Archana, P. & Pooja, S. (2013). *Pharmacosomes: An Emerging Novel Vesicular Drug Delivery System for Poorly Soluble Synthetic and Herbal Drugs.* ISRN Pharmaceutics, Article ID 348186, 10 pages.

[9] Thushara, B. D., Prasanna, R. Y, Harini, C. V, Jyotsna, T, Gowri, Y, Naga, L. M, Vivek, K. P. (2014). A perspective overview on lipospheres as lipid carrier systems. *Int. J. Pharm. Investig.*, *4(4)*, 149–155.

[10] Danielle, I, Chengan, D, Feng, L. (2017). Lipid–Drug Conjugate for Enhancing Drug Delivery. *Molecular Pharmaceutics*, *14(5)*, 1325-1338.

[11] Sandy, V, Nicolas, A, Pascal, G, Jean-Pierre, B, Patrick, S. (2011). Reverse micelle-loaded lipid nanocarriers: A novel drug delivery system for the sustained release of doxorubicin hydrochloride. *Eur. J. Pharm. Biopharm.*, *79*, 197–204.

[12] Ramezanpour, M, Leung, S. S. W., Delgado-Magnero, K. H., Bashe, B. Y. M., Thewalt, J, Tieleman, D. P. (2016). Computational and experimental approaches for investigating nanoparticle-based drug delivery systems. *BiochimicaetBiophysicaActa*, *1858*, 1688–1709.

[13] Smith, R. & Tanford, C. (1972). The critical micelle concentration of l-α-dipalmitoylphosphatidylcholine in water and water/methanol solutions. *Journal of Molecular Biology*, *67(1)*, 75–83.

[14] Brooks, B. R, Bruccoleri, R. E, Olafson, B. D, States, D. J., Swaminathan, S, Karplus, M. (1983). CHARMM: A program for macromolecular energy, minimization, and dynamics calculations. *J. Comp. Chem.*, *4(2)*, 187–217.

[15] Martínez, L, Andrade, R, Birgin, E. G, Martínez, J. M. PACKMOL: A package for building initial configurations for molecular dynamics simulations. (2009). *J. Comput. Chem.*, *30(13)*, 2157-2164.

[16] Martínez, J. M. & Martínez, L. (2003). Packing optimization for automated generation of complex system's initial configurations for molecular dynamics and docking. *J. Comput. Chem.*, *24(7)*, 819-825.

[17] Chaichian, M. & Demichev, A. (2001). Path Integrals in Physics Volume 1: Stochastic Process and Quantum Mechanics (1), Taylor & Francis.

[18] Hestenes, M. R. & Stiefel, E. (1952). Methods of Conjugate Gradients for Solving Linear Systems. *Journal of Research of the National Bureau of Standards, 49(6)*, 409-436.

[19] Dhivya, S, Sureshkumar, C, Vijayakumar, B, Rethavathi, J, Meena, C, Usha, T, Sushilkumar, M. (2018). A study of comparative modelling, simulation and molecular dynamics of CXCR3 receptor with lipid bilayer. *Journal of Biomolecular Structure and Dynamics, 36(9)*, 2361-2372.

[20] MacKay D. J. C. (1995). A free energy minimization framework for inference problems in modulo 2 arithmetic. In: Preneel B. (eds) Fast Software Encryption. FSE 1994. Lecture Notes in Computer Science, vol. 1008. Springer, Berlin, Heidelberg.

[21] Michael, L, Miriam, H, Ruth, S, Valerie, D.(1995). Potential energy function and parameters for simulations of the molecular dynamics of proteins and nucleic acids in solution. *Computer Physics Communications, 91(1–3)*, 215-231.

About the Editor

Mohammad Ashrafuzzaman, DSc
Associate Professor, Department of Biochemistry, College of Science, King Saud University, Riyadh, Saudi Arabia

Dr. Mohammad Ashrafuzzaman is a biophysicist and condensed matter scientist. He works in interdisciplinary fields using both theoretical and experimental techniques. He is passionate about trying to understand cell-based processes. Besides publishing in many reputed journals he authored two books 'Membrane Biophysics' and 'Nanoscale Biophysics of the Cell'. He is especially known for discovering a generalized charge-based membrane protein-lipid interaction mechanism. He has co-invented a US patent 'method for direct detection of lipid binding agents in membrane' to deal with drug design, drug delivery, and drug-target binding statistics.

INDEX

A

acid, 6, 8, 11, 12, 32, 33, 61, 131, 133, 134, 135, 137, 157, 180
action potential, 153
active compound, 148
acyl chain, 7, 134, 136, 142, 154
adaptation, 46, 100, 116
adrenaline, 45, 46, 51, 52, 57, 65, 76, 81, 82, 83, 84
adsorption, 20, 48, 67, 68, 75, 154
agent, 152, 182, 193
alamethicin, 13, 14, 152, 155, 181, 184
alcohol consumption, 191
amino acids, 10, 21, 124, 131, 134, 136, 137
amphipathic, 12, 18, 37, 123, 124, 132, 136, 147, 184, 195, 197
amphiphile, 155
androgens, 46, 60, 65, 68, 73, 76
animal, 17, 19, 22, 26
anion, 15, 134, 139
anionic, 10, 101, 137
antibiotic, 123, 126, 136, 140, 141, 143, 146, 147, 149, 150
anticancer drug, 171, 190, 195
antimicrobial, 124, 127, 132, 138, 145, 146, 147, 148, 153, 154, 163, 171, 180, 181, 182, 184
antitumor, 124, 138, 154
apoptosis, 31, 33, 36, 39, 171, 184
Arabidopsis thaliana, 40
arachidonic acid, 8
arithmetic, 214
atomic force, 48, 76, 80, 85, 130, 144
atomic force microscope, 48
atoms, 64, 197, 201, 202, 206, 207
ATPase, 11, 28, 47, 61, 62, 63, 65, 83, 85, 89, 90
Avogadro number, 64

B

bacillus, 124, 131, 132, 133, 134, 136, 140, 143, 144, 146, 147, 148, 149, 150
Bacillus subtilis, 144, 147, 149, 150
bacteremia, 140
bacteria, 19, 125, 126, 131, 135, 138, 140, 142, 143, 148, 149
barrier, 6, 10, 11, 24, 26, 29, 30, 111, 136, 182

bending, 5, 7, 41, 80, 82, 87, 93, 94, 95, 96, 97, 99, 101, 104, 106, 108, 109, 110, 114, 116, 117, 120, 197
bilayer, v, vii, ix, 1, 2, 3, 5, 6, 7, 9, 12, 13, 18, 20, 36, 37, 38, 39, 43, 44, 47, 52, 53, 59, 60, 63, 64, 65, 66, 75, 83, 84, 85, 88, 93, 94, 95, 96, 97, 99, 101, 103, 104, 105, 106, 108, 109, 110, 111, 112, 113, 114, 115, 116, 117, 118, 120, 121, 127, 128, 130, 135, 138, 141, 142, 147, 152, 155, 157, 172, 173, 174, 176, 177, 181, 182, 184, 185, 195, 202, 214
bilayer-cortex coupling, 94, 97, 110, 111, 115, 117
binding energies, 207
bioavailability, 190, 192, 194, 195, 196
biochemical, vii, viii, 3, 11, 38, 46, 84, 89
bioenergetic oxidative phosphorylation, 35
biological, vii, viii, ix, 1, 2, 3, 10, 14, 19, 31, 40, 89, 90, 119, 120, 123, 125, 127, 139, 140, 141, 142, 144, 145, 154, 169, 197, 210, 212
biological activities, 139
biomolecules, 1, 2, 127, 141, 178, 182, 198, 207
biophysics, vii, 3, 5, 10, 15, 36, 86, 88, 119, 121, 125, 140, 141, 215
biosurfactants, 123, 126, 131, 138, 145, 147, 148, 149
blood, 46, 47, 61, 91, 118, 119, 120, 121, 191, 194
blood vessels, 191
Boltzmann constant, 107, 108, 110
bond, 56, 57, 58, 59, 68, 134, 137, 180, 183, 197, 208
breast cancer, 87, 152, 154, 156, 159, 168, 169, 177, 180, 182, 190, 191, 192, 198, 202, 205, 212

C

calcium, 10, 15, 28, 37, 45, 86, 95
cancer, vi, 17, 87, 139, 140, 141, 151, 152, 153, 156, 159, 163, 168, 169, 175, 177, 178, 180, 182, 189, 190, 191, 192, 198, 202, 205, 212
cancer cell cytotoxicity, 152
cancer cells, 153, 169
cancer therapy, 163, 212
capacitance, 14, 17, 18, 19
capacitive, 15, 19
capsaicin, 39, 155
carbohydrates, 6, 21, 127
carboxylate, 10
carcinoma, 191, 192
cardiolipin, 7, 32, 33, 34, 36, 39, 139, 150
carrier protein, 11, 12
catecholamines, 60, 61, 88
cell, vii, viii, ix, 2, 3, 5, 11, 12, 14, 15, 16, 17, 18, 19, 20, 21, 22, 26, 31, 35, 36, 37, 38, 39, 40, 41, 44, 45, 46, 63, 66, 67, 68, 76, 77, 84, 86, 87, 88, 89, 90, 91, 95, 100, 114, 116, 117, 118, 119, 120, 121, 122, 127, 129, 133, 136, 138, 139, 146, 148, 149, 151, 153, 155, 156, 159, 168, 169, 170, 175, 177, 178, 179, 180, 182, 191, 192, 197, 215
cell culture, 45, 156
cell death, 136, 138, 169
cell line, 87, 151, 155, 156, 170, 178, 182
cell lines, 151, 155, 156, 170, 178, 182
cell membranes, vii, 2, 40, 68, 86, 89, 146, 154, 175
cell metabolism, 156
cell-penetrating peptides, 180
cellular, vii, viii, ix, 1, 2, 6, 8, 15, 16, 20, 31, 34, 36, 135, 151, 169, 178, 183, 192
ceramide, 8, 9, 33, 39, 40, 144, 172, 174, 184

chain, 7, 10, 12, 31, 89, 99, 110, 132, 133, 134, 135, 137, 143, 174
channel, 12, 13, 20, 25, 31, 36, 38, 44, 86, 87, 105, 118, 142, 145, 147, 152, 153, 162, 163, 171, 172, 173, 174, 175, 176, 178, 181, 182, 184, 185
charge, 15, 16, 17, 21, 57, 105, 127, 134, 145, 152, 154, 175, 195, 196, 215
CHARMm, 198, 208
chemical, viii, 3, 6, 10, 12, 14, 17, 20, 22, 28, 126, 127, 128, 140, 141, 160, 169, 180, 196, 202, 206
chemical properties, 6
chemical structures, 6, 202, 206
chemotherapeutic agent, 152, 154
chemotherapy, vi, 151, 152, 153, 162, 170, 173, 175, 178, 179, 182, 183, 192
chemotherapy drugs, 152, 153, 173, 178, 179
chloride, 15, 16, 87
cholesterol, 2, 3, 5, 7, 10, 19, 31, 60, 94, 95, 99, 101, 114, 115, 117, 118, 128, 129, 132, 134, 135, 144
circuit, 19, 34
classes, 3, 6, 135, 148, 172, 176
clinical examination, 191
clustering, 93, 101, 115, 117, 170
colchicine, vi, 151, 152, 153, 154, 155, 156, 158, 159, 168, 169, 170, 177, 178, 179, 182, 183, 184
collaboration, viii
complex interactions, 123, 178, 182
composition, 7, 21, 31, 32, 33, 34, 36, 37, 46, 60, 61, 94, 125, 127, 128, 129, 130, 132, 134, 135, 136, 138, 142, 172, 196
compounds, 124, 126, 127, 131, 134, 135, 139, 145, 149, 156, 160, 171, 184, 194
compression, 57, 69, 70, 72, 73, 77, 79, 82, 83, 86, 96, 98, 104
computational biology, vii, 211
concentration, 7, 11, 15, 16, 45, 48, 49, 50, 51, 52, 53, 54, 55, 61, 62, 63, 64, 65, 81, 83, 85, 95, 120, 132, 135, 138, 149, 152, 159, 160, 163, 169, 170, 194, 196, 198, 213
conductance, 14, 152, 160, 161, 162, 163, 165, 171, 173, 174, 186
conjugategradient(CG), 199, 200, 203, 204
conjugation, 154, 196
constituents, 15, 20, 31, 32, 94, 97, 100, 111, 152, 160
contour, 25, 103, 107
cortex, 94, 96, 97, 99, 102, 106, 108, 109, 110, 111, 112, 113, 115, 117
cortisol, 46, 51, 52, 56, 57, 58, 59, 74, 81, 82, 83, 84, 88
Coulomb, 15, 18, 104, 152, 155, 176, 177
crassa, 32, 33, 34, 36
critical value, 46, 63, 75
cross-fertilization, vii
culture medium, 156
curvature, 7, 39, 40, 67, 68, 95, 96, 99, 101, 104, 108, 109, 114, 127, 135, 154, 181
cycles, 100, 116, 158
cylindrical, 7, 8, 14, 21, 80, 172, 173
cytoplasm, 22, 23, 26, 29, 178
cytoskeletal, 3, 4, 5, 41, 86, 87, 119
cytosolic, 8, 12
cytotoxicity, 151, 152, 154, 155, 156, 160, 168, 169, 170, 177, 178, 183, 192

D

daptomycin, 132, 136, 137, 140, 142, 143, 144, 146, 147, 148, 149, 150
decane, 18, 157
decay, 163
defects, 13, 119, 155, 171, 174
defence, 124
deformation, 63, 72, 73, 78, 79, 84, 95, 104, 105, 109, 130, 174
dehydration, 58, 68
depolarization, 136

depolymerization, 45
derivatives, 111, 134, 139, 144, 169, 180, 182, 183, 184
detection, 170, 183
deviation, 202, 205
dielectric, 17, 18, 21, 40, 118
dielectric constant, 18
diffusion, 3, 11, 20, 30, 93, 94, 95, 99, 101, 110, 114, 115, 117, 119, 121, 122, 172, 181
Dipalmitoylphosphatidylcholine (DOPC), 18, 198, 212
disease, 22, 27, 28, 31, 37, 121, 192
diseases, 31, 35, 140, 152, 153, 178
displacement, 69, 70, 72, 73, 74, 79
dissection, 24
domain, 4, 5, 28, 31, 56, 69, 110, 129, 143, 154, 165
drug, vi, ix, 140, 148, 151, 153, 154, 155, 157, 158, 159, 160, 161, 162, 163, 164, 165, 167, 168, 169, 170, 174, 175, 176, 177, 178, 183, 189, 190, 192, 193, 194, 195, 196, 197, 198, 199, 201, 202, 203, 204, 205, 206, 207, 208, 209, 210, 211, 212, 213, 215
drug delivery, 190, 192, 193, 194, 195, 196, 198, 201, 202, 204, 205, 209, 210, 211, 212, 213
drug-delivery, 190
dynamics, vi, 21, 31, 40, 90, 97, 118, 121, 128, 146, 153, 158, 181, 184, 185, 189, 190, 197, 199, 200, 201, 202, 204, 206, 209, 210, 211, 212, 213, 214
dynamics and simulation, 190

E

EDTA, 61, 159
electric field, 16, 172, 174, 176
electrical, 14, 15, 16, 17, 19, 118, 154, 160, 177
electrical properties, 19
electrogenic, 30
electron, viii, 24, 25, 35, 130
electron microscopy, 35
electron paramagnetic resonance, 130
electrophysiology, viii, 18, 152, 153, 155, 157, 160, 170
electrostatic energy, 152, 202, 206, 207, 208, 211
endophytic bacteria, 125, 126, 142, 143, 149
endoplasmic reticulum, 26, 27, 28, 33
endosomal, 154, 180
endothelial, 153
energy minimization, 199, 200
energy parameters, 202, 206, 208, 210
envelope, 22, 23, 24, 25, 26, 27, 28, 36, 37, 38, 41, 45, 75, 148, 151, 179
environment, 1, 2, 10, 12, 17, 127, 151, 155, 176
environmental conditions, 100, 135, 144, 190, 198
enzymatic, 12, 62, 131
enzymatic activity, 62
enzyme, 45, 61, 62, 63, 64, 65, 83, 168
equilibrium, 5, 71, 75, 80, 102, 105, 129, 130, 153, 158, 165, 207, 209
erythrocyte membranes, 44, 46, 47, 48, 49, 53, 55, 56, 57, 61, 62, 63, 65, 78, 83, 85, 86, 88, 89, 122
erythrocytes, 38, 45, 46, 48, 56, 60, 66, 68, 75, 76, 80, 81, 89, 90, 112, 120, 122
ethanesulfonic acid, 157
ethanol, 53
eukaryotic, 6, 7, 8, 10, 44
experimental condition, 101, 163, 172, 187
extracellular, 1, 2, 4, 5, 10, 13, 16, 18, 21, 44, 86
extracts, 133

F

families, 86, 124, 140
fatty acids, 10, 11, 137, 144, 197
fencing, 3
Fengycin, 135, 144, 150
fermentation, 142
filament, 28, 45, 94, 97, 99, 103, 106, 107, 115, 117
financial, 141, 192
financial support, 141
flexibility, 97, 103, 116, 117
fluctuations, 96, 98, 110, 111, 117, 163, 197, 205
fluid, 3, 6, 7, 9, 10, 21, 38, 39, 40, 94, 128, 148
fluid -mosaic, 3, 21
fluorescence, 49, 52, 53, 54, 85, 89, 119, 129, 130, 149
F-MMM, 3, 4, 5
fMRI, viii
food, 124, 141, 148
Food and Drug Administration, 153
force, 12, 30, 34, 43, 44, 45, 47, 49, 64, 68, 69, 70, 83, 106, 107, 120, 149, 158, 183, 197, 198, 202, 208
formation, 3, 13, 28, 31, 56, 57, 58, 59, 68, 74, 76, 79, 83, 84, 96, 97, 98, 100, 105, 110, 114, 115, 117, 121, 125, 127, 131, 132, 135, 136, 139, 142, 143, 147, 151, 155, 156, 157, 169, 170, 174, 175, 178, 179, 181, 182, 194, 197
formula, 54, 80, 152
free energy, 38, 51, 52, 56, 95, 104, 108, 109, 214
freedom, 136, 158, 201

G

galactosylceramide, 9
gel, 9, 10, 74, 83, 94, 128, 135, 179
gene expression, 45, 46
genetic defect, 91
geometry, 7, 8, 134, 145
glucose, 6
glucosylceramide, 8, 9
glycerol, 10
glycoprotein, 3
glycoproteins, 3, 4, 21
gramicidin, 13, 38, 120, 152, 155, 180, 181, 185
growth, vii, 57, 64, 65, 83, 133, 139, 163, 183, 192

H

harmonic potential, 158
headgroup, 7, 128, 133
health, 1, 2, 22, 31, 35, 156, 192
health condition, 31, 35
height, 13, 48, 75, 78, 80, 167
hemoglobin, 66, 96, 100
hexagonal lattice, 90
hierarchical, viii, 22
hormone, 46, 49, 50, 51, 52, 54, 55, 56, 58, 59, 62, 63, 81, 82, 85
human, 35, 38, 46, 86, 87, 90, 114, 118, 119, 120, 121, 138, 154, 170, 184
hydrocarbon, 10, 172, 174
hydrocarbons, vii, 3, 19, 177
hydrolysis, 11, 33, 61, 63, 65, 103, 131, 196
hydrophilic, vii, 5, 82, 89, 124, 131, 155, 193, 195, 197
hydrophobic, vii, 5, 7, 9, 12, 14, 18, 21, 38, 56, 57, 59, 63, 68, 69, 72, 76, 96, 115, 120, 122, 124, 127, 131, 155, 165, 172, 173, 175, 177, 193, 195, 196, 197
hypercholesterolemia, 45
hypothesis, 26, 47, 65, 83, 170
hypoxia, 65, 66

I

immunoglobulin, 28
in vitro, 145, 156, 160, 177, 180
in vivo, 6, 38, 135, 138, 145, 179
independent variable, 111
information, iv, vii, ix, x, 3, 4, 14, 21, 130,
    154, 185, 197
infrared spectroscopy, 130
inhibition, 135, 138, 156, 179
insilico, 190, 198, 202
insulating, 14, 18
integrity, 96, 100, 115, 116, 117
interdependence, 163
interface, 127, 136, 144, 174, 193, 195, 197
interference, 24, 170
intracellular, 1, 2, 10, 13, 16, 18, 40, 45, 100
ion channel, vi, 5, 16, 20, 44, 45, 127, 139,
    145, 151, 152, 153, 155, 171, 181, 183
ion channels, 5, 16, 20, 44, 127, 139, 145,
    152, 153, 155, 171, 181, 183
ion pore, 151, 152, 155, 156, 157, 171, 173,
    176, 178, 182
ionic, 10, 94, 102, 103, 105, 117, 135, 138,
    149
ions, 11, 15, 16, 19, 20, 22, 30, 45, 63, 136,
    138, 160, 163, 198
ischemia, 37
issues, x, 15, 17, 157, 164
Iturin A, 134

J

joints, 76
Josephson junction, 185

K

K^+, 11, 15, 16, 17, 45, 46, 47, 61, 62, 63, 65,
    66, 83, 85, 87, 88, 90, 136

kinetics, 147, 163, 177
kurstakin, 132, 133

L

lactate dehydrogenase, 156
lamellar, 9, 110, 128
lamina, 26, 27
Langevin dynamics, 158
lateral motion, 115
lead, 56, 93, 100, 101, 110, 114, 115, 117,
    152, 154, 158, 171
leukocyte, 153
lifetime, 13, 38, 61, 96, 100, 115, 116, 117,
    152, 161
light, ix, 5, 8, 34, 47, 93, 149, 154, 171, 178
lipid bilayer bending state, 94, 95, 97, 114
lipid bilayer membrane, 5, 18, 20, 151, 152,
    155, 173, 176, 177, 178, 183, 184, 186
lipid drug delivery vehicle, 193
Lipid-Based Drug Delivery Systems, 192,
    212
lipopeptides, v, 123, 124, 126, 131, 133,
    134, 138, 140, 141, 142, 143, 144, 145,
    146, 148, 150
lipoproteins, 131
liposome, 130, 154, 212
liposomes, 125, 130, 154, 180
localization, 25, 28, 35, 198
lupus erythematosus, 179
lysis, 47, 122, 143, 146, 180
lysosomal, 154, 171, 180, 184

M

machinery, 40
macromolecules, 6, 202, 206
mammalian cells, 9
mass, 25, 133, 141, 164, 165, 166, 191, 201
materials, vii, ix, 2, 10, 11, 20, 22, 36, 84,
    91

Index

MCF-7, 156, 159
MD Simulation, 158
measurements, 14, 18, 48, 50, 51, 53, 54, 63, 86, 130, 178, 184
mechanical, iv, v, 3, 5, 43, 44, 46, 47, 56, 60, 63, 64, 66, 69, 70, 71, 76, 77, 83, 84, 85, 87, 90, 95, 123, 125, 127, 130, 147, 152, 155, 183, 194, 198
mechanical properties, 3, 5, 123, 130, 155
mechanical stress, 43, 44, 46, 47, 60, 66, 69, 70, 71, 76, 77, 83, 84, 85
media, 17, 18, 67, 121, 139, 159, 160, 169
Mediterranean fever, 153
membrane permeability, 168
membranes, ix, 1, 2, 6, 8, 10, 11, 14, 18, 20, 22, 24, 26, 27, 29, 30, 31, 35, 36, 37, 38, 39, 40, 43, 45, 46, 47, 53, 54, 55, 56, 59, 60, 62, 63, 65, 66, 73, 75, 76, 80, 83, 84, 85, 89, 90, 118, 119, 120, 123, 125, 126, 127, 130, 132, 134, 135, 137, 142, 143, 144,145, 146, 147, 151, 153, 154, 155, 157, 160, 161, 170, 174, 175, 176, 177, 178, 180, 181, 182, 183, 184, 186, 198
mesoscale, 22
messengers, 8
metabolites, 123, 124, 140, 141, 146
microbial community, 126
microdissection, 24
microemulsion, 194
micromanipulation, 24
micromolar, 155
microorganisms, 124, 125, 131, 138
microscopy, 35, 43, 47, 48, 49, 76, 80, 85, 118, 130, 137, 144
microtubule, 86, 87, 156, 179, 180, 183
microviscosity, 44, 46, 47, 53, 54, 55, 60, 63, 64, 65, 66, 69, 75, 83, 85, 88
mitochondria, 11, 29, 30, 31, 32, 33, 34, 35, 37, 40, 152, 156
mobility, 4, 35, 95, 102, 107, 114, 119
model system, 129
models, 3, 47, 66, 125, 143

modifications, 21
modulus, 70, 79, 99, 104, 106, 107, 108, 109, 110, 112, 113, 120, 130
molecular, vi, vii, 6, 7, 8, 26, 36, 38, 118, 119, 120, 121, 122, 126, 127, 130, 132, 134, 139, 141, 142, 143, 144, 146, 147, 152, 153, 154, 155, 156, 165, 174, 176, 178, 181, 182, 183, 185, 189, 197, 198, 199, 200, 201, 206, 208, 209, 212, 213, 214
molecular dynamics, 146, 153, 181, 185, 197, 199, 200, 201, 206, 209, 212, 213, 214
Molecular dynamics, 185, 197, 201, 209
molecular structure, 155
molecules, 6, 7, 8, 11, 13, 18, 21, 22, 26, 30, 37, 48, 52, 56, 60, 67, 68, 69, 72, 73, 74, 80, 94, 95, 96, 97, 99, 101, 102, 103, 104, 105, 106, 107, 108, 110, 114, 115, 116, 117, 121, 123, 125, 127, 131, 135, 136, 139, 141, 154, 155, 158, 160, 164, 165, 170, 174,176, 179, 183, 184, 195, 196, 197, 199, 206
molybdenum, 61
monolayer, 12, 13, 14, 109, 130
monomer, 14, 163, 173
monomers, 13, 14, 53, 163, 173
morphology, 39, 41, 43, 47, 48, 49, 60, 76, 85, 149
MTT, 156, 159, 160, 168, 169, 182, 183
mucosa, 169
multidimensional, 184
multipass, 12
mutation, 190, 192
mycorrhiza, 125

N

Na^+, 11, 15, 16, 17, 45, 46, 47, 61, 62, 63, 65, 66, 83, 85, 87, 88, 89, 90
NaCl, 61, 157

NADPH, 156
nanocrystals, 46, 75, 84, 88
nanometer, 5, 197
nanoparticles, 46, 53, 75, 76, 198
National Bureau of Standards, 214
National Institutes of Health, 156
needle, 24
neurospora, 32, 33, 34, 36
neutral, 7, 154, 156, 175, 183
noradrenaline, 45, 47, 48, 49, 50, 51, 52, 53, 54, 55, 57, 60, 62, 67, 68, 76, 81, 82, 83, 84, 85
nuclear, v, ix, 1, 2, 22, 23, 24, 25, 26, 27, 28, 36, 37, 38, 40, 41, 45, 87, 130, 149, 151, 153, 179
nuclear magnetic resonance, 130, 149
nuclear membrane, ix, 1, 2, 22, 25, 26, 154
nucleoporin, 25
nutrients, 2, 20

O

occlusion, 46, 85
off-target, 154, 171
Ohmic, 19
optimization, 116, 202, 214
organ, 7
organelle, 36
organelles, 6, 11
osmotic pressure, 67, 68, 100
osmotic stress, 95, 121
oxidation, 195
oxide nanoparticles, 89

P

PACKMOL, 198, 199, 213
parallel, 18, 19, 24, 63, 111
pathogens, 123, 126, 136, 140, 145, 149, 182
pathways, 26, 87

penicillin, 136
peptides, 11, 13, 58, 124, 127, 130, 146, 147, 149, 153, 163, 171, 175, 178, 182, 184, 206
perinuclear, 23, 25, 27
permeability, 6, 10, 34, 85, 136, 168, 190, 192
permeation, 22
permission, iv, 29
pH, 10, 29, 47, 50, 53, 61, 115, 135, 138, 152, 157, 158, 162, 165, 174, 186, 187, 198
pharmaceutical, 124, 136, 140, 211
phase, viii, 7, 10, 39, 74, 83, 95, 98, 114, 119, 127, 128, 129, 130, 132, 135, 140, 143, 151, 154, 155, 164, 180, 187, 193, 194, 195, 197, 201, 204
phase transitions, 154
phosphate, 8, 10, 33, 47, 48, 50, 51, 53, 58, 65, 158, 165
phosphatidic, 6, 32, 33
phosphatidylcholine, 6, 31, 32, 33, 132, 143, 145, 157, 172
phosphatidylinositol, 6, 12, 31, 32, 33
phosphatydyleserine, 157, 172
phosphoethanolamine, 124, 157, 172
phosphoinositides, 8
phospholipid, 1, 2, 6, 8, 10, 31, 32, 33, 34, 36, 53, 58, 59, 118, 135, 139, 143, 144, 146, 152, 154, 158, 162, 172, 176, 177, 180, 181, 185, 196
phospholipids, 3, 6, 7, 10, 19, 31, 32, 33, 34, 56, 57, 58, 59, 60, 68, 73, 74, 135, 176, 177, 181
phosphorylation, 30, 35, 38, 97, 103
physical, vii, viii, 1, 2, 6, 84, 88, 91, 120, 127, 129, 142, 153, 154, 155, 175, 176, 178, 180, 195
physical interaction, 155, 175, 176
physical properties, 6, 127, 154, 180
physics, vii, viii

Index

physiological, vii, 1, 2, 3, 13, 14, 23, 85, 178, 179
plant, 22, 24, 28, 37, 38, 40, 124, 125, 138, 141, 150, 156
plasma, v, ix, 1, 2, 3, 6, 7, 11, 17, 20, 21, 22, 38, 39, 88, 119, 120, 122, 135, 179, 196
plasma membrane, ix, 3, 7, 11, 17, 20, 21, 22, 88, 119, 120, 122, 135
platelet, 180
polar, 7, 10, 12, 21, 128, 133, 135, 139, 143
polymerization, 44, 45, 153, 156, 169
Polymyxin B, 138
pore, v, 23, 26, 38, 40, 123, 131, 132, 136, 137, 139, 142, 147, 149, 151, 153, 155, 161, 163, 165, 170, 171, 172, 173, 174, 175, 176, 178, 181, 187
positive hydrophobic mismatch effects, 93, 94, 95, 96, 101, 104, 117
potential, 15, 16, 17, 29, 30, 34, 41, 65, 66, 132, 138, 139, 140, 141, 143, 148, 152, 153, 156, 157, 158, 171, 197, 199, 200, 202, 203, 204, 206, 207, 208, 210, 211, 214
potential energy, 34, 197, 199, 200, 202, 203, 204, 206, 207, 208, 210, 211, 214
probability, 25, 152, 177
process, 11, 24, 30, 63, 75, 84, 100, 103, 115, 116, 117, 158, 194, 198, 199, 200, 201, 202, 203, 204, 209, 210, 211, 214
proliferation, 139, 183, 191
protein, 4, 5, 12, 15, 20, 21, 22, 26, 27, 29, 30, 31, 32, 33, 34, 37, 38, 39, 40, 48, 49, 50, 51, 52, 53, 54, 55, 56, 57, 60, 61, 62, 63, 66, 69, 70, 73, 74, 76, 81, 82, 83, 84, 89, 90, 93, 94, 95, 96, 97, 100, 101, 104, 105, 115, 117, 118, 119, 120, 127, 136, 142, 146, 148, 149, 153, 155, 156, 171, 174, 176, 181, 197, 202, 215
protein-protein interactions, 95, 101
proteins, vii, 2, 3, 4, 5, 7, 8, 10, 11, 12, 13, 19, 20, 21, 22, 24, 26, 27, 28, 30, 31, 32, 34, 35, 36, 37, 38, 39, 44, 46, 48, 51, 52, 53, 54, 55, 56, 57, 58, 59, 60, 61, 62, 63, 68, 69, 70, 73, 74, 77, 82, 85, 91, 95, 105, 118, 119, 120, 122, 125, 127, 147, 153, 163, 171, 197, 202, 206, 214
proton, 29, 30, 34
protonation, 10
pseudocystalline, 9
pseudomonas, 132, 138, 140, 142, 145

Q

quartz, 50, 53, 88

R

radiation, ix, 192
radius, 66, 75, 104, 106, 108, 130, 171
radius of gyration, 106
rearrangement of lipids, 94, 95
receptor, 12, 28, 35, 87, 192, 214
receptors, 40, 45, 46, 52, 76, 151, 198
relaxation, 99, 100, 184
relevance, 124, 128, 130
resistance, 19, 94, 125, 126, 141, 182
resting, 13, 16, 17, 26, 153, 173, 174
reticulum, 26, 27, 28
reverse micelles, 190, 192, 197, 198, 200, 201, 202, 203, 204, 205, 206, 207, 208, 212
Reverse micelles, 197
room temperature, 7, 130, 193, 209
root, 28, 37, 125, 209

S

scattering, viii, ix, 118, 181
screened Coulomb potential, 177
second generation, 183
sedimentation, 47
seed, 209

sex hormones, 51
shape, vii, xi, 20, 43, 60, 66, 67, 68, 75, 76, 95, 162, 171, 196, 198
sialic acid, 9
signal, 28, 39, 40, 122
signal transduction, 122
simulation, vi, 118, 155, 158, 164, 165, 166, 176, 177, 185, 189, 190, 197, 200, 201, 202, 204, 206, 208, 209, 210, 211, 214
single crystals, 91
single-pass, 12
skeleton, 22, 89, 90, 122
sodium, 11, 15, 16, 17, 39, 44, 45, 63, 86, 87
software, 157, 158
solubility, 154, 157, 164, 190, 192, 194
solution, 6, 48, 51, 61, 66, 94, 96, 100, 102, 113, 117, 149, 158, 159, 160, 172, 214
solvation, 165
species, 10, 31, 134
spectroscopy, ix, 130, 149, 184
sphingolipid, 8, 33, 41
sphingomyelin, 7, 8, 9, 32, 95, 124
sphingosine, 8
sphingosylphosphorylcholine, 8
stability, 43, 152, 157, 160, 162, 168, 172, 175, 177, 190, 195, 196, 198, 202, 204, 205, 206, 207, 208, 209, 211
stable complex, 202, 204, 206, 208
state, 10, 17, 20, 51, 55, 56, 74, 93, 94, 95, 96, 97, 100, 101, 105, 106, 110, 114, 115, 116, 117, 135, 163, 184
states, 9, 31, 91, 99, 129, 163, 171
Steepestdescent(SD), 32, 142, 149, 150, 199, 200, 203, 204
stem cells, 83
sterols, 6, 10, 145
stimulation, 44
stimulus, 16, 198
STM, viii
stock, 157, 159, 164, 169
stoichiometry, 148, 174

storage, 30, 98, 99, 109, 112
stress, 7, 44, 46, 51, 65, 68, 71, 73, 75, 76, 77, 78, 79, 80, 84, 88, 89, 110, 111, 112
stretching, 56, 58, 100, 105, 130
strong interaction, 154
structural changes, 59, 66, 84, 93, 97, 98, 99, 100, 101, 104, 108, 110, 113, 137
structural characteristics, 154
structural transformations, 47
structural transitions, 55, 56, 57, 58, 61, 83, 84, 85
structural transitions in biomembranes, 55, 56, 83, 85
structural variation, 210
structure, 3, 5, 6, 10, 13, 18, 20, 21, 26, 39, 40, 46, 47, 51, 56, 57, 58, 59, 60, 68, 73, 76, 83, 84, 85, 87, 88, 89, 90, 96, 105, 110, 115, 116, 124, 128, 132, 133, 136, 140, 146, 156, 168, 174, 175, 180, 184, 197, 198, 202, 209
subgel, 9
substitution, 59, 79
substrate, 78
sulfate, 10, 52
SUN protein, 24, 28
surface area, 8, 128, 194, 195
surface chemistry, 198
surface tension, 128, 130, 139
surfactants, 141, 145, 190, 193, 197
surfactin, 124, 132, 133, 134, 140, 143, 144, 146, 148, 149
swelling, 96, 100, 101, 105, 112, 113, 114, 116, 117, 121
synthesis, 22, 30, 33, 136, 138, 143, 145, 148
synthetic, 153, 213
syringomycin E, 139, 142, 145, 146, 147, 148, 149

T

taxol, 151, 153, 154, 155, 158, 180, 183, 185
temperature, 7, 48, 51, 53, 94, 107, 110, 127, 128, 129, 130, 131, 135, 138, 158, 193, 194, 198, 201, 204, 205, 209, 211, 212
tension, 70, 73, 74, 80, 82, 83, 106, 127, 129, 130, 139, 194
testosterone, 46, 51, 52, 58, 63, 76, 81, 83, 84
therapeutic, 153, 170
therapeutic effect, 170
thermodynamic properties, 198
thermodynamic stability, 204
thiocolchicoside, 151, 155
tissue, 191, 193
titania, 47
tobacco, 191
torsinA, 27
total energy, 198
toxicity, 151, 153, 169, 192, 196
translocase, 35
transmembrane, 11, 12, 14, 21, 40, 73, 74, 77, 119, 122, 152, 157, 172
transport, vii, 1, 2, 10, 11, 20, 21, 26, 30, 31, 32, 34, 36, 41, 44, 89, 154, 157, 179
treatment, 45, 140, 146, 149, 164, 192, 198
Trypsin, 159
tubulin, 151, 153, 155, 156, 169, 179, 180, 182, 183
tumor progression, 154

types of breast cancer, 191

U

underlying mechanisms, 178
unsaturated, 7, 9

V

validation, 190
vancomycin, 136
vector, 70, 72, 73, 74, 79
velocity, 64, 65, 99, 109, 112
vesicle, 6, 26, 129, 130, 143
vomiting, 169

W

Washington, 180
water, 56, 69, 73, 101, 128, 136, 145, 147, 158, 195, 196, 212, 213
worldwide, 190

X

X-axis, 169

Y

Y-axis, 169
yeast, 26, 135, 149
yield, 61, 204

Related Nova Publications

CALMODULIN: STRUCTURE, MECHANISMS AND FUNCTIONS

EDITOR: Vahid Ohme

SERIES: Cell Biology Research Progress

BOOK DESCRIPTION: In *Calmodulin: Structure, Mechanisms and Functions*, the authors consider small and poorly-studied groups of plant calcium-dependent protein kinases that directly interact with calmodulin molecules.

SOFTCOVER ISBN: 978-1-53614-948-7
RETAIL PRICE: $82

FLAGELLA AND CILIA: TYPES, STRUCTURE AND FUNCTIONS

EDITOR: Rustem E. Uzbekov

SERIES: Cell Biology Research Progress

BOOK DESCRIPTION: Motility is an inherent property of living organisms, both unicellular and multicellular. One of the principal mechanisms of cell motility is the use of peculiar biological engines – flagella and cilia. These types of movers already appear in prokaryotic cells. However, despite the similar function, bacteria flagellum and eukaryote flagella have fundamentally different structures.

SOFTCOVER ISBN: 978-1-53614-333-1
RETAIL PRICE: $95

To see a complete list of Nova publications, please visit our website at www.novapublishers.com

Related Nova Publications

CHLOROPLASTS AND CYTOPLASM: STRUCTURE AND FUNCTIONS

EDITORS: Carl Dejesus and Lourdes Trask

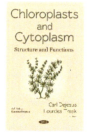

SERIES: Cell Biology Research Progress

BOOK DESCRIPTION: In this collection, the authors begin by presenting one approach for studying the cyclosis-mediated transmission of regulatory signals along the characean internode by illuminating a cell part with a narrow light beam and to observe the changes of modulated chlorophyll fluorescence occurring in response to this photostimulus at a fixed distance on the downstream side of the light spot.

HARDCOVER ISBN: 978-1-53614-127-6
RETAIL PRICE: $160

MYOSIN: BIOSYNTHESIS, CLASSES AND FUNCTION

EDITOR: David Broadbent

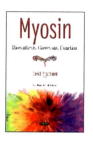

SERIES: Cell Biology Research Progress

BOOK DESCRIPTION: *Myosin: Biosynthesis, Classes and Function* opens with a discussion on class I myosins, the most varied members of the myosin superfamily and a remarkable group of molecular motor proteins that move actin filaments and produce force.

SOFTCOVER ISBN: 978-1-53613-817-7
RETAIL PRICE: $95

To see a complete list of Nova publications, please visit our website at www.novapublishers.com